DENSIFICATION OF METAL POWDERS DURING SINTERING

STUDIES IN SOVIET SCIENCE

DENSIFICATION OF METAL POWDERS DURING SINTERING

V. A. Ivensen
All-Union Scientific Research Institute for Hard Alloys
Moscow, USSR

Translated from Russian by
Eric Renner

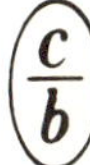 **CONSULTANTS BUREAU • NEW YORK – LONDON • 1973**

Vladislav Aleksandrovich Ivensen was born in Moscow in 1908. In 1931 he was graduated from the Institute of Fine Chemical Technology as a specialist in the technology of rare metals. His first work centered on the alumino-thermal method of obtaining carbon-free metals. Since 1936 his work has been concerned with the technology of hard alloys. He has been the director of one of the laboratories of the All-Union Scientific Research Institute of Hard Alloys since 1947. In the same year he received the degree of Candidate of Technical Sciences for his work on the theory of sintering.

The original Russian text, published for Metallurgiya Press in Moscow in 1971, has been corrected by the author for the present edition. This translation is published under an agreement with Mezhdunarodnaya Kniga, the Soviet book export agency.

KINETIKA UPLOTNENIYA METALLICHESKIKH POROSHKOV PRI SPEKANII
V. A. Ivensen

КИНЕТИКА УПЛОТНЕНИЯ МЕТАЛЛИЧЕСКИХ ПОРОШКОВ ПРИ СПЕКАНИИ
ИВЕНСЕН Владислав Александрович

Library of Congress Catalog Card Number 72-94822
ISBN 0-306-10881-X

A Division of Plenum Publishing Corporation
227 West 17th Street, New York, N. Y. 10011

United Kingdom edition published by Consultants Bureau, London
A Division of Plenum Publishing Company, Ltd.
Davis House (4th Floor), 8 Scrubs Lane, Harlesden, NW10 6SE, London, England

Printed in the United States of America

Preface

Sintering of powder metal compacts is one of the basic operations in powder metallurgy. The useful properties of a machine part are obtained after considerable densification of the sintered material. Although the mechanical properties of the part depend on other structural factors besides porosity, porosity is the main factor. Usually, the practical problem in sintering is to obtain a part with the desired or permissible porosity. Thus, knowledge of the laws governing densification and its final result is necessary to control this process in the production of powder metal parts.

The laws governing densification are also important for a more exact physical theory of sintering, which is still in the initial stages of its development. Such processes as the change in the density of lattice defects and the flow of crystalline substances during sintering have not yet received a complete physical interpretation. Analysis of the laws of sintering may provide additional material for more complete phenomenological characteristics of these processes that will be useful for further development of theoretical concepts of the flow of imperfect crystals under small loads.

Although a substantial amount of experimental material has been accumulated, generalizations are still difficult.

Many of the published reports deal with "models" of sintered bodies (beads or wires sintered to each other or to a flat plate, and also solid bodies with fine drilled holes). Such experiments have certain advantages (the possibility of exact quantitative calculation of the change in the area of contact or change in the diameter of pores), but do not make it possible to evaluate the basic processes that occur in sintering of real powders, the densification of which is closely connected to the nonequilibrium condition of the crystal lattice of the metal. On the other hand, the investigations of the sintering of compacts made so far do not give a very clear idea of the kinetics of densification, which would result if the process could be isolated from the effect of outside factors (the gas pressure in closed pores, for example). In some cases densification occurs in a relatively short length of time, which has frequently led to erroneous or exceedingly rough estimates of densification as a function of time (as will be shown later). The empirical formulas that have been published satisfactorily describe the course of densification during sintering for a long period but not in the beginning of sintering.

In this author's opinion, the formulation of hypotheses and theories of sintering based on strict physical concepts must be preceded by a fairly complete phenomenological study of the process. The schematic and often unpersuasive nature of the theoretical concepts is due precisely to the fact that the theoretical formulations are based on an incomplete phenomenological description of the process.

The author believes that the present condition of the theory of sintering justifies a return to phenomenological investigations of the process in order to reveal the laws and, if possible, a quantitative description of densification during sintering. In solving theoretical problems in sintering one cannot bypass the empirical laws if they are fairly reliable, encompass a wide range of original materials and sintering conditions, and are expressed by simple mathematical functions. For this reason the examination of the kinetics of densification during sintering will begin with a description of the basic, most general laws of densification of powder compacts during sintering.

Let us mention some characteristic features of the terminology used in this work. Although the terms "reduction in the volume of pores" and "densification" are not synonymous, the author sometimes substitutes one for the other. Both processes are the same in essence, and the substitution of "densification" for "reduction in the volume of pores" makes it possible to shorten the text without interfering with the clarity of the presentation. The author takes this liberty in order to avoid tedious repetition of "reduction in the volume of pores."

For easier reference, the notation that is used in several chapters is presented at the beginning of the book. For convenience, the equations are numbered by the following system: The first number is the number of the chapter; the second number is the numerical order of the equation in the chapter.

Where no other source is given, the experimental data used in this work were obtained at the All-Union Scientific-Research Institute of Hard Alloys by N. V. Baranova and L. P. Usol'tseva under the direction of the author. A number of calculations, including the value of function F_T in Table 27, were made by V. A. Fal'kovskii. The electron microscopic studies (Chapter VII) were made by N. V. Baranova, N. P. Vasil'eva, N. F. Koval'skaya, and T. A. Sultanyan.

Contents

Notation

l	linear dimension of the body
Δl	change in dimension during sintering
v_p	volume of pores in compact before sintering, cm^3/g
v_s	volume of pores after sintering, cm^3/g
v	relative volume of pores, expressed as v_s/v_p
v_{in}	relative volume of pores at beginning of isothermal sintering (i.e., after constant sintering temperature is reached)
v_0	relative volume of pores at nominal beginning of isothermal sintering (at time τ_0)
V	volume of porous body in absolute units, cm^3
V_p	volume of porous body before sintering
V_s	volume of porous body after sintering
P	porosity expressed as volume percent of porous body
P_p	porosity before sintering
P_s	porosity after sintering
d	density of porous body, g/cm^3
d_p	density of compact before sintering
d_s	density after sintering
d_c	density of solid (compact) substance (without pores)
D	relative density
D_p	relative density before sintering
D_s	relative density after sintering
S	surface of interconnected pores per unit mass of porous body, cm^3/g
S_p	surface of pores before sintering
S_s	surface of pores after sintering
G	gas permeability (volume of air reduced to 760 mm Hg passing through a porous body with a cross section of 1 cm^2

and thickness of 1 cm in 1 min at some constant difference in pressure, given in the text)

G_p gas permeability of compact

G_s gas permeability after sintering

N concentration of defects (this value is used together with factors *a* and *b* and without them; it has no dimensions)

aN the kinetic characteristic of the concentration of defects determining the flow of an imperfect crystal, h^{-1}

aN_{in} the relative concentration (or kinetic characteristic of concentration) of defects at the time isothermal sintering actually begins

aN_0 the kinetic characteristic of the concentration of defects in the original powder

E_a the activation energy of the elimination of defects, cal/g-atom

E_b the activation energy of flow associated with lattice defects, cal/g-atom

α the average rate of increase in temperature to the beginning of isothermal sintering, deg/h

T absolute sintering temperature

T_m absolute melting point

τ isothermal sintering time, h

τ_{in} the time from the beginning of heating to the actual beginning of isothermal sintering

τ_0 the time from the beginning of heating to the nominal beginning of isothermal sintering (corresponds to the time coordinate of the point of intersection of the sloping and horizontal lines of the idealized temperature graph, Fig. 40)

Chapter I

Laws Governing the Relationship between the Initial and Final Densities of Sintered Bodies

A review of the laws governing densification during sintering should begin with the constant specific reduction in the volume of pores, which was established twenty years ago [1]. When porous bodies compacted from a given powder at different pressures, and differing in initial porosity, are sintered under the same conditions, the volume of pores is the same in all bodies in most cases. If v_p and v_s are the volumes of pores before and after sintering, then v_s/v_p = const (on condition that densification is not disturbed by associated processes that affect the volume of the sintered body).

This law merits attention for two reasons. On one hand, the densification process as a function of time can be investigated at the same initial density, since the course of densification described by the specific reduction in the volume of pores is the same for any initial density (within the limits in which this law holds true). On the other hand, determining the range of the initial density in which the specific reduction in the volume of pores is the same makes it possible to determine the conditions in which the densification process occurs in the "purest" form – without substantial distortion due to other processes that cause expansion of the body.

In almost all cases where the specific reduction in the volume of pores is not constant the reason for it can be found.

TABLE 1. Values of v_s/v_p, P_s/P_p, and $\Delta\Phi$* for Nickel Bodies With Different Initial Densities Sintered Under Identical Conditions†

d_p, g/cm³	d_s, g/cm³	v_s/v_p	P_s/P_p	$\Delta\Phi$
3.43	6 04	0,291	0.514	0,442
3.66	6.30	0,283	0.486	0.391
3,95	6,52	0.285	0,470	0.330
4,12	6,65	0.285	0,459	0.298
4.39	6.84	0.284	0.444	0 251
4.55	6.95	0.279	0.435	0.229
4,79	7,09	0,287	0,426	0,196
4,91	7,18	0.285	0,416	0,181
5.20	7,36	0.283	0.400	0,149
5,42	7.46	0,288	0,391	0,122

*$\Delta\Phi$ is a function of the densities before and after sintering, which, according to Skorokhod, should have a constant value at different d_p.
†Powder obtained by dissociation of carbonyl nickel. Sintered at 850°C for 60 min.

First, let us examine the case where the law holds in almost ideal fashion. The specific reduction in the volume of pores is constant in sintering of copper, nickel, iron, silver, and other powders when the powdered metals are obtained by reduction of oxides at moderately high temperatures (of order 0.5-0.8T_{melt}). The specific reduction in the volume of pores is especially constant in sintering of carbonyl nickel powder obtained by dissociation of gaseous nickel carbonyl. The particles of carbonyl powder have a regular spherical shape. The values of v_s/v_p are quite constant for bodies of this metal over a wide range of initial densities (Table 1).

Let us keep in mind that v_s/v_p is the ratio of the absolute volume of pores before sintering v_p and after sintering v_s and that this ratio can be found from the density before sintering d_p, the density after sintering d_s, and the density of the solid (compact) metal d_c by means of the following formula

$$\frac{v_s}{v_p} = \frac{d_p(d_c - d_s)}{d_s(d_c - d_p)}. \qquad \text{(I-1)}$$

When bodies with different d_p compacted from powders obtained by reduction of oxides are sintered under the same conditions the variation of v_s/v_p is somewhat larger, although in most cases the value of v_s/v_p differs from the average value no more than ±5%. From the experimental data so far obtained on the change in the volume of pores before and after sintering it follows that the values of v_s/v_p are far more often constant than not. For example, in sintering powders of various origins the value of v_s/v_p for bodies with different initial densities sintered under the same conditions differed no more than ±5% from the average value in 52 tests out of 58 for copper powders, 19 tests out of 26 for nickel powders, 9 tests out of 14 for iron powders, and 59 out of 67 tests for tungsten carbide powders (these data refer only to tests in which v_s/v_p was constant in a fairly wide range of initial densities: In each test d_p differed by more than $0.15d_c$.

In recent years there have been few quantitative calculations of the relationship between porosity (or density) before and after sintering. In 1948, Bal'shin [2] showed that for bodies with different initial densities the ratio of the porosity before and after sintering, expressed in percent of the volume of the body, is also approximately constant. However, more detailed analysis showed that the constant value is the ratio of the absolute volumes of pores and not the ratio of the porosities P_s/P_p, where P_s and P_p are expressed in percent of the volume of the body. In all cases where constant values of v_s/v_p are found experimentally, i.e., when the value falls within limits of ±5% in the range of initial densities larger than $0.15d_c$, the variation of P_s/P_p is considerably greater than ±5%. This can be seen in Table 1, where the values of P_s/P_p are given in addition to v_s/v_p for carbonyl nickel powder. Since the value of v_s/v_p is constant in most cases of sintering compacts, it is evident that the constancy of this value is the basic relationship between the volumes of pores before and after sintering (and thus also the densities before and after sintering [3]).

In 1961, Skorokhod [4] examined the variation of densification with time, which is also the source of a certain variation in density (or porosity) before and after sintering.

Skorokhod attempted to derive a kinetic equation of densification taking into account the effect of the specific concentration of pores on the macroscopic ductility of the sintered mass. A similar attempt was made earlier by Mackenzie and Shuttleworth [5], although their calculations referred only to systems with small volumes of closed pores. Based on a system of calculations proposed by Frenkel' (equating the energies scattered in viscous flow to the change in the energy of the system with reduction of the surface of pores), Skorokhod derived a similar equation for the balance of energy. This equation takes into account the effect on the value of the dissipated energy of the change in the effective coefficient of viscosity (the second coefficient of viscosity entering into the equation of the hydrodynamics of viscous liquids), depending on the value of the porosity [6]. In its final form (after separation of the variables and integrating) he obtained an equation from which it follows that after a given sintering time some complex function of the change in the volume of pores should have a constant value regardless of the initial volume of pores:

$$\Delta\Phi = \Phi_s - \Phi_p = -\frac{3}{2}\cdot\frac{\sigma}{r}\int_0^{\tau}\frac{d\tau}{\eta}, \tag{I-2}$$

where σ is the surface tension, r is the radius of the particles, η is the coefficient of viscosity, and τ is time.

Since

$$\Phi = -\frac{1}{2(1-P)} + \frac{1}{3}\ln P - \frac{3}{4}\ln(1-P) + \frac{5}{12}(3-P), \tag{I-3}$$

the corresponding values of porosity before and after sintering can be found from d_s and d_p and, substituting these values into Eq. (I-3) twice, one can find the difference: $\Phi_s - \Phi_p = \Delta\Phi$. By this means it is not difficult to plot the variation of $\Delta\Phi$ with d_p.

According to Skorokhod, the value of $\Delta\Phi$ should depend only on the sintering conditions (for a given powder). With constant sintering conditions the value of $\Delta\Phi$ should be constant for a body of any initial density. However, this was not confirmed experimentally.

Using values of d_p and d_s from any experiment in which v_s/v_p is constant in a fairly wide range of d_p, we obtain values of $\Delta\Phi$ that differ substantially for different d_p, as was shown for carbonyl nickel in Table 1. The value of $\Delta\Phi$ changes rapidly with increasing d_p. Hence it follows that the variation of d_s with d_p cannot be described by Skorokhod's equation. The apparent agreement with the experiments resulted from the use of a value of the density before and after sintering that was in the range of the initial densities, taking in both the region in which v_s/v_p is constant and the region in which forces inducing expansion of the body are active (gas pressure in closed pores). As will be shown below, in these experiments the change in density depended on the combined influence of compacting and expansion processes. Studies of the change in the volume of pores under such sintering conditions do not permit observation of a regular variation with time or initial density and often lead to erroneous conclusions. We shall take up this question later.

On the basis of Mackenzie's and Skorokhod's assumptions, the constant value of the reduction in the volume of pores could have the following semiphenomenological explanation: With increasing density of the body the average pore size decreases, and consequently the curvature of the surface bounding the pore increases, which leads to an increase in the negative capillary pressure around the pore. However, the reduction of the porosity leads simultaneously to an increase in the effective (macroscopic) coefficient of viscosity. An increase of the effective viscosity compensates for the effect of increasing negative capillary pressure around the pore, and therefore the specific rate of "healing" of pores does not change. Regretfully, experiments do not confirm this outwardly rational explanation. The value of Skorokhod's function decreases rapidly with increasing initial density, which indicates either more change in the actual effective viscosity than follows from the purely rheological assumptions or that the average size of pores with a complex shape does not depend completely on the value of capillary forces inducing reduction of the surface and volume of pores.*

Although there is still no agreement on the reason for the constant value of the specific reduction in the volume of pores, this law must be taken into account in investigating the kinetics of densification. In the overwhelming majority of cases the experiments show that with connecting pores in bodies with different initial porosities the volume of pores is reduced to an equal degree under the same sintering conditions and that the change in the volume of pores with time follows the same law [1]. If the specific reduction in the volume of pores is constant for a given powdered material, then this relationship holds for any sintering conditions. The only change is in the range of the original density of compacts for which the specific reduction in the volume of pores is constant.

*In connection with another model of porosity, Skorokhod recently found [7] that P_s/P_p should be independent of d_p. As was shown above, this conclusion is also unconfirmed by experiments (see Table 1).

Chapter II

Conditions for Observing Densification Process in Pure Form

In plots of v_s/v_p vs. d_p one observes both random and systematic deviations from constant values of v_s/v_p. The first may be due to nonidentical sintering conditions of bodies with different initial porosities and other accidental circumstances. Systematic deviations that are consistently observed in sintering of bodies from a given powder are usually closely connected with the structural characteristics. In most cases the reason for the deviation from a constant value of the specific reduction in volume of pores can be determined.

As was shown in [1, 8], at relatively small d_p the reduction in the volume of pores is almost constant for powders of ductile metals, but at some value of density (later called the critical density) the value of v_s/v_p tends to increase with increasing d_p and often reaches unity or higher. The increase of v_s/v_p when the critical density is exceeded is due to the development of processes causing expansion of the body. As was indicated above, in the range of initial densities below the critical density the value of v_s/v_p is not completely constant. In some cases v_s/v_p increases slowly with increasing d_p. Sometimes the value of v_s/v_p decreases gradually with increasing d_p.

The specific reduction in the volume of pores during sintering of metal powders obtained by reduction of oxides with hydrogen is usually constant in the case where the reducing temperature is fairly high. At relatively low reducing temperatures (400-500°C for copper and nickel, for example) the value of v_s/v_p

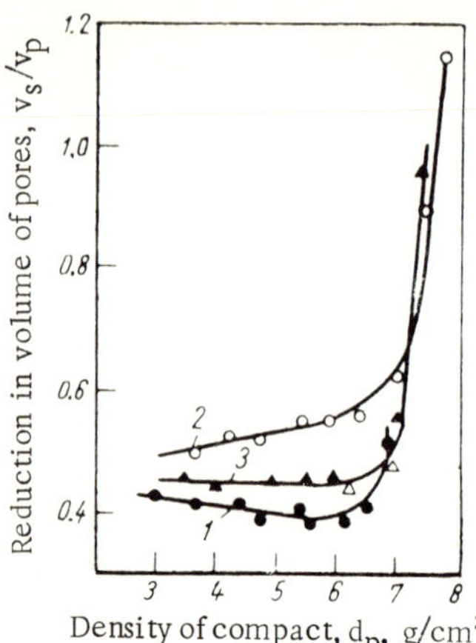

Fig. 1. Variation of v_s/v_p with d_p for different copper powders. 1) Powder with apparent density 1.32 g/cm^3 obtained by reduction of oxides with hydrogen at 450°C; sintered at 720°C; 2) mixture of copper powders of different dispersity obtained by reduction of oxides at 500 and 800°C (the latter additionally calcined at 900°C), sintered at 900°C; 3) powder with apparent density 2.06 g/cm^3 obtained by reduction of oxides at 750°C, sintered at 880°C.

decreases with increasing d_p and then increases sharply with the development of processes inducing expansion (Fig. 1, curve 1).

It is well known that particles of powders obtained by reducing oxides with hydrogen have a spongy texture, which is only slightly evident (or absent) at high reducing temperatures but is pronounced at relatively low reducing temperatures. It is probable that the specific rate of reduction of small internal pores (pores within the particles) is higher than the rate of reduction of pores (voids) between particles. Since the percentage of pores within particles in the total volume of pores increases with increasing initial density of the body (increasing compacting pressures), the rate of reduction in the total volume of pores during sintering increases somewhat with increasing initial densities.

Other explanations are also possible. Since the shape of the particles obtained by reduction at low temperatures is irregular and complex, and increase of the compacting pressure increases not only the area of contacts but also the number of contacts between particles, which leads to a substantial difference in the geometry of the connected, channel-like pores in weakly and strongly compacted bodies. In some manner the pronounced spongy structure of particles results in some small but systematic deviation of v_s/v_p from a constant value.

A deviation of the specific reduction in the volume of pores from a constant value in the opposite direction (increasing v_s/v_p

with increasing d_p in the range below critical density) is observed in sintering mixed powders obtained by reduction of oxides at temperatures differing considerably. Figure 1 (curve 2) shows the variation of v_s/v_p with d_p for bodies compacted from equal amounts of copper powders obtained by reduction of oxides at 500 and 800°C. In this case the increase of v_s/v_p with increasing initial density is evidently due to the fact that the probability of direct contact between "low-activity" particles obtained at higher reducing temperatures increases with increasing compacting pressures. In cases where low-activity particles are separated by high-activity particles the effect of the latter on the overall densification process increases and vice versa, with direct contact (or close proximity) of the denser low-activity particles of high-temperature powder the densification process depends to a considerable extent on sintering of low-activity particles, which occurs at a slower rate. In some cases a similar deviation of v_s/v_p from a constant value is observed for powders differing in origin, including those obtained by reduction of oxides. It is quite probable that these powders differ greatly in the size and properties of particles. Evidently, the more uniform the size of the original particles, with a regular shape and the smallest internal porosity, the more constant the value of the specific reduction in volume of pores. This is confirmed by the constant value of v_s/v_p in a wide range of d_p for sintered carbonyl nickel powder with spherical particles (see Table 1).

A systematic deviation from a constant value of the specific reduction in volume of pores occurs in the region of high initial densities exceeding the critical density (Figs. 2 and 3). The rapid

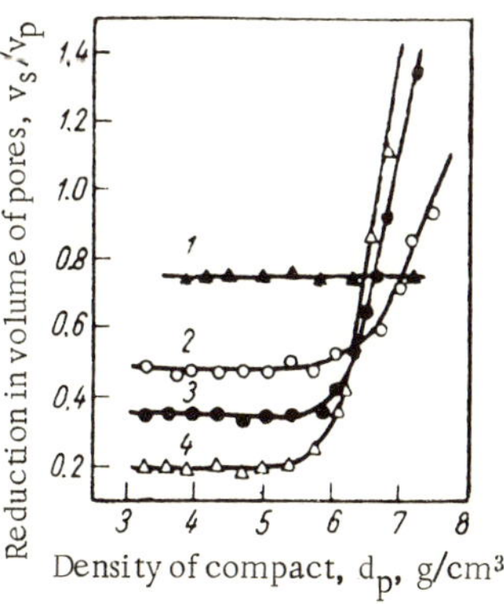

Fig. 2. Variation of v_s/v_p with d_p at different sintering temperatures of copper powder with apparent density 1.85 g/cm³. 1) 640°C; 2) 750°C; 3) 850°C; 4) 950°C. Sintering time 30 min.

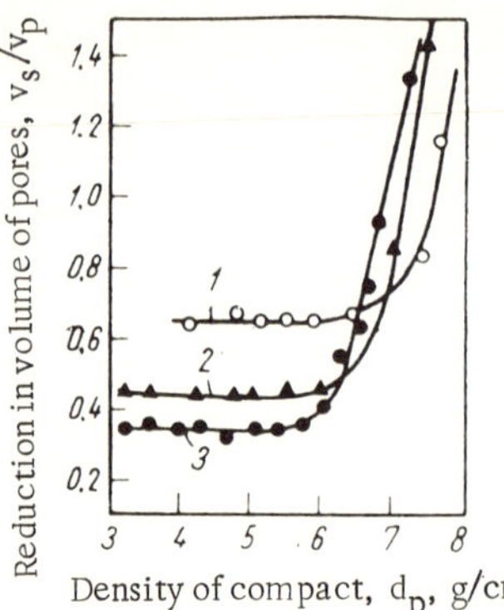

Fig. 3. Variation of v_s/v_p with d_p for copper powders obtained by reduction of oxides at different temperatures. Compact sintered 30 min at 850°C. 1) Reduced at 780°C, (apparent density 2.46 g/cm^3); 2) 640°C (apparent density 1.85 g/cm^3); 3) 520°C (apparent density 1.45 g/cm^3).

increase in the value of v_s/v_p with increasing d_p is due to the development of processes inducing expansion of the sintered body. The rapid increase of v_s/v_p with increasing d_p in sintering of relatively ductile material (many pure metals) compacted at high pressures may give values of v_s/v_p higher than unity, i.e., an increase in the volume of pores as compared with the compact. An investigation of the structure of pores in such bodies leaves no doubt that the main process causing expansion of the body under these conditions is the evolution of sorbed gases in closed pores of the sintered body.

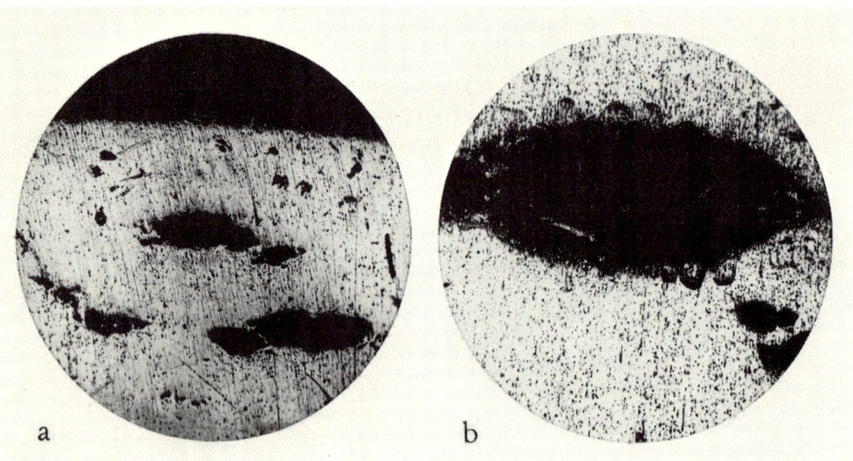

Fig. 4. Small (a) and large (b) lenticular pores in a copper sample that increased in volume after sintering (density before sintering, 8.08 g/cm^3, sintering temperature 850°C) (50×).

At short sintering times the large pores in bodies undergoing expansion are often lens-shaped, with the long section of the lens perpendicular to the direction in which the compacting pressure was applied (Fig. 4). This orientation of lenticular pores is evidently due to the direction of maximum elastic expansion after removal of the compacting pressure. Elastic expansion primarily in the direction of the compacting pressure reduces the strength of contacts, which during subsequent sintering favors breaking of the material under the influence of the gas pressure in the direction perpendicular to the application of pressure. It is also possible that the shape of the pores is significant in compacts of ductile metals obtained under high pressure. It is as if the pores were flattened out, with the length in the direction perpendicular to the application of pressure far larger than in the direction of the application of pressure. Therefore the stresses created by gas pressure in the direction of pressing the compact are considerably higher than in other directions, which is responsible for the respective orientation of breaks in the material. After prolonged sintering, the large pores resulting from the evolution of sorbed gases become rounded (Fig. 5).

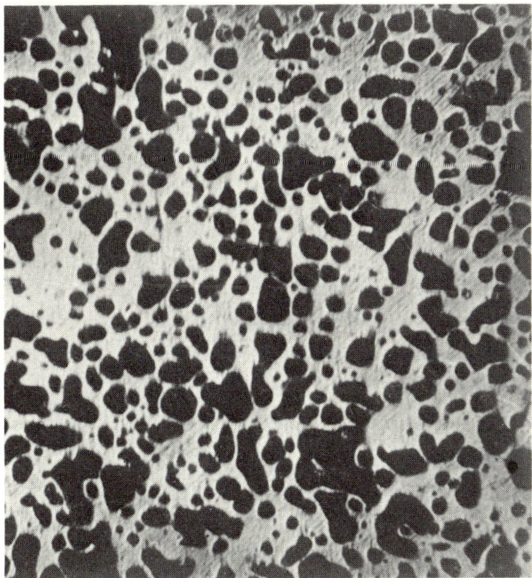

Fig. 5. Large pores in sample of silver powder after prolonged sintering (500 h) at 800°C (100×).

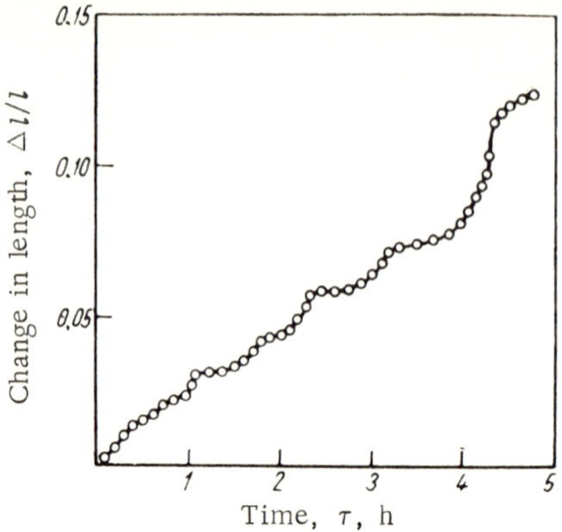

Fig. 6. Dilatometric curve in the direction of the compacting pressure for nickel with a high initial density (d_p = 7.58 g/cm³).

The fact that the change in the volume of the body at the beginning of sintering under the influence of expansion processes results from intermittent breaks in the material with formation of large pores is confirmed by the dilatometric curves obtained from measurements of expansion on a sintered body in the direction of the compacting pressure (Fig. 6). The expansion periodically accelerates and slows down, each acceleration induced by new breaking of contacts with formation of a large cavity. Unlike the densification curves (with "pure" densification), the expansion curves reflect the erratic character of the process, including the initial rapid formation of voids in different parts of the sintered body, which gradually changes into the slower process of expansion and rounding of pores. At small compacting pressures and moderately high initial densities the densification and expansion processes coexist. This is not obvious where densification occurs in the beginning of sintering and then changes to expansion. It must be assumed that the change from densification to expansion is due to the occurrence of closed pores. The joining of channel-like pores and their separation into closed sections creates conditions for development of a substantial gas pressure that inhibits reduction of the volume of the pore.

The evolution of sorbed gases from the sintered body evidently continues over a long period of time. In some experiments with prolonged sintering of silver compacts densification was replaced by expansion after sintering for a very long time, in the course of which the body contracted. Expansion of the body began after sintering for 75 h at 800°C. The minimal value of v_s/v_p was 0.319. During subsequent sintering the body continued to expand and at the end (400 h) the volume of pores corresponded to a value of $v_s/v_p = 0.955$.

A change from densification to expansion is often observed in sintering bodies of different powders with an initial density corresponding to the initial increase in the value of v_s/v_p with an increase of v_p. Quite often this is observed at a fairly high sintering temperature (Fig. 7).

The gas pressure in closed pores does not invariably lead to expansion of the body. Often it slows down the densification process, disrupting its normal course. In plots of v_s/v_p vs. d_p this is manifest in an increase of v_s/v_p with increasing d_p. The fact that the disruption of the constant value of v_s/v_p is due to the gas pressure in closed pores is confirmed by the following fact: With a continuous increase of d_p the initial growth of v_s/v_p usually coincides with the disappearance or a substantial reduction of the percentage of interconnected pores in the body.

As an example, let us consider the results from sintering of copper powder with an apparent density of 2.22 g/cm^3 obtained by reduction of oxides. The initial density varied from 3.31 to 8.18 g/cm^3 and the compacting pressure from 380 to 10,000 kg/cm^2. The value of v_s/v_p was approximately constant in the range of initial densities from 3.31 to 6.20 g/cm^3. With further increase of the initial density the value of v_s/v_p increased rapidly (Fig. 8). Both the total volume of pores and the percentage of the volume

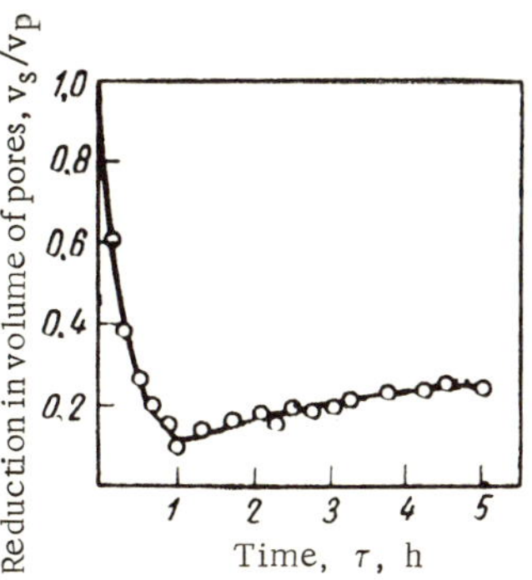

Fig. 7. Change from densification to expansion during sintering (850°C) of active nickel powder with $d_p = 4.65$ g/cm^3.

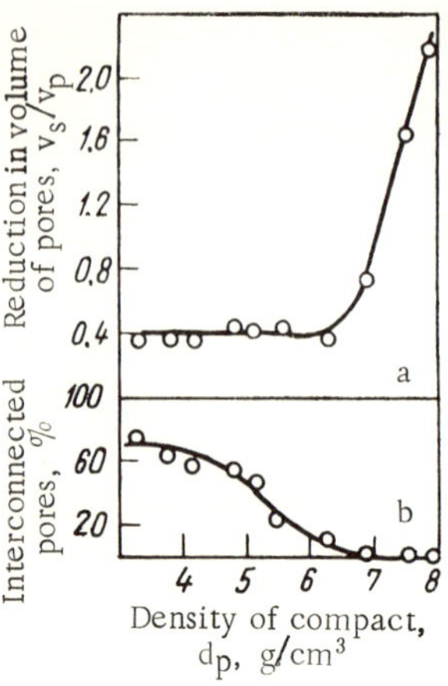

Fig. 8. Comparison of reduction in volume of pores (a) and percentage of interconnected pores after sintering (b) with density of compacts (copper powder sintered at 875°C).

of pores filled on submersion of the body in melted paraffin (interconnected pores) were determined. It was found that the value of v_s/v_p begins to increase only after the body ceases (or almost ceases) to absorb paraffin, which indicates almost complete disappearance of interconnected pores, ensuring unimpeded elimination of gases (see Fig. 8). The occurrence of barriers to the elimination of sorbed gases is the main reason for the development of the expansion process.

The relationship between expansion and the gas pressure in closed pores was confirmed in the work of Pines and Sirenko [9], where it was found that compacting of copper powder in vacuum sharply reduces the effect of expansion during subsequent sintering even in the case of a high initial density.

Microscopic examination showed quite different types of porosity in bodies sintered under conditions of pure densification (constant value of v_s/v_p) and under conditions where the expansion process is dominant (i.e., with $v_s/v_p > 1$). In the first case one observed fine pores more or less evenly distributed throughout the section. With intensive development of the expansion process after sintering one observed the previously described lenticular or spherical pores and sometimes pores of more complex shape formed from several joined lenticular pores or large pores.

Together, these observations indicate that in the range of densities corresponding to a constant value of v_s/v_p there are no direct or indirect indications of the process of expansion during isothermal sintering. The process of expansion develops and coexists with the process of densification only in the range of higher initial densities, in which the value of v_s/v_p increases. Hence, it follows that plotting the variation of v_s/v_p with d_p is a convenient means of determining the sintering conditions in which densification occurs in its purest form. During sintering of real powders the densification process can be observed without distortion in most cases only with interconnected pores, providing free outlet of sorbed gases from the body. The indication of undisturbed densification is the constant value of v_s/v_p with a change of initial density.

The region of undistorted densification (or the range of initial densities corresponding to undistorted densification) varies substantially with the sintering conditions or the properties of the powders. With an increase of the sintering temperature the range of constant values of v_s/v_p becomes narrower (see Fig. 2). With equal sintering conditions the range of undisturbed densification may be wider or narrower, depending on the properties of the original powders (see Fig. 3). It is readily noted that disturbance of the normal course of densification, manifest in an increase of v_s/v_p, occurs at small d_p if the sintering conditions or properties of the powder ensure the formation of closed pores at low densities of compacts. This is favored both by an increase of sintering temperature and the use of more dispersed powders, since both accelerate densification. Along with this, the more probable the formation of closed pores in the sintered body, the more easily normal densification is disrupted. The formation of closed pores is intimately connected with the ductility of the powdered substance (closed pores are formed at low pressures during compacting of powders of ductile metals). During sintering of tungsten carbide and titanium compacts there is almost no plastic deformation under normal conditions, and v_s/v_p generally does not increase (Fig. 9), since interconnected pores are retained in sintered bodies even at very high compacting pressures (up to 20,000 kg/cm^2).

The lack of any analysis of sintering conditions from the viewpoint of external effects or internal reasons for disruption of the densification process has more than once led to erroneous conclusions concerning the validity or invalidity of some relation-

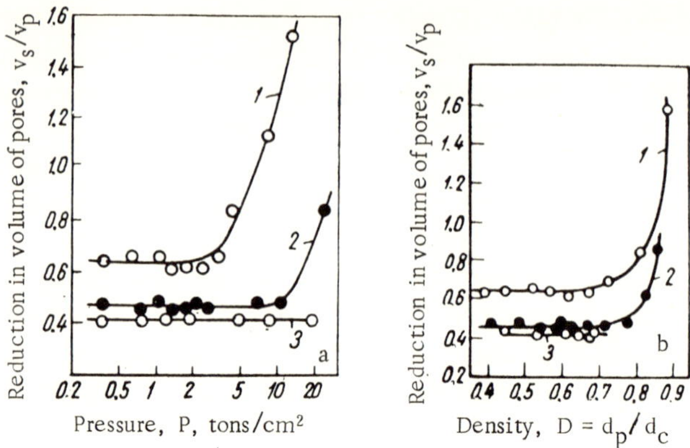

Fig. 9. Variation of v_s/v_p with compacting pressure (a) and relative density of compact (b) pressed from powders differing in ductility. 1) Copper, sintered at 800°C; 2) iron, sintered at 1200°C; 3) tungsten carbide, sintered at 1550°C.

ship. This can be illustrated by attempts to use Skorokhod's experimental data from [1] to determine the fitness of his kinetic equation without accounting for the expansion process in the range of high initial densities. Presenting plots of P_s vs. P_p (P_p and P_s are the porosity before and after sintering) for nickel powders from experimental data [1], Skorokhod did not note that the experimental points first depart from the curve matching his equation and then (after development of the expansion process) again approach this curve, which created the illusion of satisfactory agreement between the curve described by the equation and the experimental data ([4], Fig. 3, curve 2). However, if the plot is compared with the plot of v_s/v_p vs. P_p, making it possible to determine the range of undisturbed densification, where v_s/v_p is constant, then it is not difficult to establish that in this range the experimental data characterizing the variation of P_s with P_p do not coincide with the calculations (the experimental curve intersects the theoretical curve at a substantial angle – see Fig. 10*).

Skorokhod considered the small but systematic deviation of the data on copper powder from the theoretical curve to be un-

*It should also be noted that the correlation in [4] was derived from experiments where v_s/v_p was constant in a limited range of d_p and that it was not very consistent. If the calculations are made for a more typical case with a wide range of densities where v_s/v_p is constant, then the disagreement between the calculations and the experimental data becomes far more evident.

important [4]. However, the densification in this case was relatively small. If the experimental and theoretical values had been compared for an experiment where $v_s/v_p < 0.5$ then the deviation would have been as large as in the experiment with nickel powders cited above. If a fairly wide range of values of P_p is used then the experimental curve for any experiment with copper powder (with $v_s/v_p < 0.5$) will intersect the theoretical curve twice – in the region of undisturbed densification and in the region where the processes of densification and expansion coexist (as is shown for nickel powder in Fig. 10). In the intermediate region, in passing from pure densification to expansion one can always select sections where the slopes of the theoretical and experimental curves coincide (in a narrow range of P_p), although it can obviously not be regarded as confirmation of the validity of the premise of the theoretical curve.

The systematic deviation of the experimental data from the theoretical curve in the range where v_s/v_p is constant reflects the absence of any constant value of function $\Delta\Phi$ with changes of P_p or d_p, as was demonstrated earlier for the case of carbonyl nickel powder (see Table 1 for densification in the pure form).

It should be noted that in the range where densification and expansion processes coexist any variation of the density and porosity (with dispersity of the powder, initial density, sintering temperature and time) can be assumed to be quite different, with sometimes unexpected shapes, and it is impossible to draw general conclusions from individual observations under these conditions.

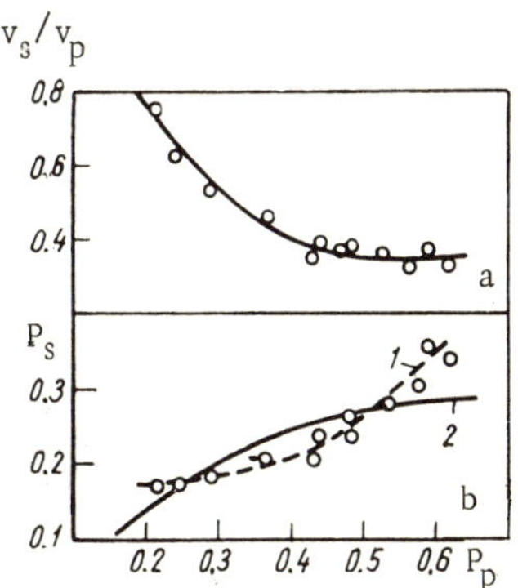

Fig. 10. Variation of v_s/v_p with P_p (a) and P_s/P_p (b) from tests used by Skorokhod [4] to compare experimental data with theoretical curves. 1) Variation of P_s with P_p plotted from data in [1] (nickel powders, sintering temperature 790°C); 2) same, calculated from Skorokhod's equation.

A single-valued variation of density with sintering conditions, repeated on different powders, can be established only with observation of the densification process in the pure form, i.e., under sintering conditions where the specific reduction in the volume of pores is constant. These conditions frequently, although not always, coincide with retention of interconnected pores up to the end of sintering. Sometimes the deviation from a constant value of v_s/v_p begins with an increase of closed porosity to 20-40% and the presence of a still substantial percentage of interconnected pores. On the other hand, in the experiments of Grube and Schlecht [153], used by the present author [1], a constant value of v_s/v_p was found in sintering of molybdenum powder with a density up to $0.97d_c$, i.e., under conditions where the through (interconnected) pores completely or almost completely disappear after sintering. Therefore, determining the percentage of through pores in compacts is not a reliable means of determining the conditions for undisturbed densification. These conditions should be established, as was indicated above, by plotting the variation of v_s/v_p with d_p for given sintering conditions and by finding limit values of d_p where v_s/v_p is constant.

Although the expansion of a body during sintering is most often due to evolution of sorbed gases in closed pores, the gas pressure is not the only reason for the expansion of sintered bodies. According to Bal'shin, the expansion of a body instead of ordinary densification was found in sintering of coarse copper powder under conditions excluding the effect from the evolution of sorbed gases [2]. The same observation was made in sintering compacts of fine iron filings.

In considering the processes capable, along with densification, of affecting the volume of the sintered body one should take into account the possible development of "thermoplastic" or "elastic-plastic" aftereffects (in Fridman's terminology [10]). This phenomenon, due to the anisotropic elastic and plastic characteristics of crystals composing the polycrystalline substance, is evident in the fact that the deformed body changes its shape during heating, usually in the direction opposed to the application of force during deformation. The thermoplastic aftereffect undoubtedly develops to some extent during heating of compacted metal powders.

Dilatometric curves of densification made by the differential method, which excludes the effect of thermal expansion, frequently show expansion of the body in the first minute of heating in the direction opposite that of deformation (compacting). This expansion is rapidly replaced with shrinkage of the body due to the development of densification.

Judging from the character of the isothermal densification curves and from the range of temperatures in which the thermoplastic aftereffect develops (350-400°C for copper, 400-600°C for iron), this process is practically completed during heating to a given sintering temperature and therefore has no effect on the densification curves at constant temperature.

Among other possible reasons for the disruption of pure densification are changes in the volume of the sintered body and the densification rate due to allotropic transformations. The general laws of densification should be investigated with metal powders not undergoing phase transformations in the range of sintering temperatures. As for the effect of oxide films, it differs substantially, depending on whether the oxide was reduced in the atmosphere of the sintering furnace or whether it is resistant and is retained throughout the sintering period.

Oxidation of the surface of particles of copper, nickel, and other metal powders, the oxides of which are easily reduced in hydrogen (before compacting), as when a compact is exposed to an oxidizing atmosphere, changes the value of v_s/v_p somewhat with sintering in hydrogen but does not affect the constant value of the reduction in the volume of pores and has little effect on the range of d_p in which it is constant. However, the addition of oxides that are difficult to reduce (Al_2O_3, Cr_2O_3, etc.) has a substantial effect on the variation of v_s/v_p with d_p, causing a deviation of v_s/v_p from constant (particularly an increase of v_s/v_p with increasing d_p).

Films of reducible oxides are rapidly removed in the beginning of sintering in a hydrogen atmosphere and do not interfere with the densification process, which occurs in fairly pure form with isothermal sintering conditions.

Chapter III

Volume of Pores in Relation to Isothermal Sintering Time

An important stage in studying the mechanism of a complex physical process is the investigation of its kinetic characteristics by observing the changes with time of the parameters characterizing the development of the process. Although measuring the reduction of linear dimensions or the volume of a sintered body with time presents no experimental difficulties, the investigation of the kinetics of densification of powders is far from complete. Most investigations have been limited to qualitative or semiquantitative descriptions of the process, and only in recent years have empirical formulas been presented with constants that have no clear physical or phenomenological interpretation.

Before the publication of work concerning the kinetics of densification a number of investigators had concluded that several elementary processes probably coexisted during sintering. The possibility of describing any process as an elementary process obeying the laws of chemical kinetics is determined by investigating the temperature dependence of the time necessary for a certain stage of completion of the process (characterized by some level of the properties). If the phenomenon is based on a single elementary process with a specific activation energy then the relationship between the logarithm of time and the reciprocal of absolute temperature should be linear.

The process of densification during sintering was analyzed in this manner as early as 1950 by Huttig [11]. However, his conclusions were not in good agreement with the experimental results. Attaching no importance to the small but systematic curvature of lines on plots of log τ vs. $1/T$ for several powders, Huttig considered that densification during sintering could be regarded as an elementary process. In a similar analysis soon after that, Duwez [12] observed the systematic deviation log τ vs. $1/T$ from a linear relationship, which indicated the complex nature of densification.

At the present time there is no doubt that there are several interrelated processes that directly or indirectly affect densification during sintering. The question is only whether they coexist (to a fairly discernible extent) throughout the sintering process or whether their relative influence on the sintering process changes so much during sintering that one process can be said to replace another. In other words, it is not clear whether the process of sintering single-phase bodies should be regarded as a complex process with several stages and a distinct difference in the importance of elementary processes (at different stages) or a single process in which the contribution of the main elementary processes can be observed both in the beginning and after sintering for some time. At the same time, the elementary processes themselves and their complex interactions in the process of sintering have not been studied in detail. A simple enumeration of the possible elementary processes with no analysis of their actual part in sintering of real powders (and not models) explains little.

All these questions can be fully answered only by detailed studies of the laws of the kinetics of densification during sintering of powdered substances in combination with other investigations characterizing other changes occurring in a sintered body.

The data in the literature concerning densification with time are inadequate for analyzing the kinetic characteristics of this process. The existing kinetic equations frequently use the values of linear or volume contraction or "parameters" of densification that are ratios of the densities of the sintered, compacted, and solid metal which, as will be shown below, less accurately and less graphically characterize the process than the specific reduction in the volume of pores. In those cases where the kinetic equation describes the change in the volume of pores it proves to be suitable only for the initial section of the curve or, on the contrary, suitable except for the initial period of isothermal sintering.

An analysis of the different kinetic equations will be presented after a description of experimental data characterizing the variation of v_s/v_p with τ for different powders and sintering conditions.

Since a study of the density after sintering in relation to the density of the compact revealed a regular relationship – a constant value of v_s/v_p – sintering as a function of time could also be expressed by this same ratio.

As we said above, with undisturbed densification under different sintering conditions the change in the volume of pores per unit of initial volume of pores is the same in bodies with different initial densities. The value of v_s/v_p remains constant at practically any sintering time [1]. This means that the variation of v_s/v_p with τ coincides for bodies with different densities. Thus, the variation of the density of all bodies of a given powder with the sintering time, in the absence of disrupting influences, is determined by the common relationship $v_s/v_p = f(\tau)$, which can be presented graphically or by an empirical formula. Hence the task of plotting the variation of v_s/v_p with τ and finding empirical formulas capable of describing this relationship under different sintering conditions.

Experimental studies of the densification process can be made conveniently by means of a dilatometer with rapid increase of the temperature to the beginning of isothermal sintering (for this purpose the furnace, heated to a given temperature, is placed near the dilatometer). In order to eliminate errors due to thermal expansion, the differential method is used with a specified difference in the length of the porous and solid samples of metal investigated. In our investigations the variation in temperature did not normally exceed ±5°C. Sintering was conducted in hydrogen.

The results of dilatometric measurements were used to find the variation of the specific reduction in the volume of pores, v_s/v_p, with sintering time. To calculate v_s/v_p from the value found for Δl (shortening of the length), the specific reduction in the volume of the sintered body, V_s/V_p, was determined first:

$$\frac{V_s}{V_p} = \frac{(l - \Delta l)(a - \beta\Delta l)(b - \gamma\Delta l)}{V_p}, \qquad \text{(III-1)}$$

where l, a, and b are, respectively, the length, thickness, and width of the rectangular compacts before sintering, Δl is the shortening of the length determined in the dilatometer,

$$\beta = \frac{\Delta a_{fi}}{\Delta l_{fi}} \text{ and } \gamma = \frac{\Delta b_{fi}}{\Delta l_{fi}},$$

where Δl_{fi}, Δa_{fi}, and Δb_{fi} are the final contractions (changes in linear dimensions) in length, width, and thickness determined from measurements of the sample after sintering.

With this method of calculating the shrinkage (from changes in the dimensions of the sample) there is some error because of the smallness of the changes in the ratio of contraction in three directions. However, special experiments where the shrinkage was determined by means of direct measurements of the volume and by calculation, using formula (III-1), for several values of sintering time showed that the maximum error in v_s/v_p did not exceed 3%.

The specific reduction in the volume of pores was found from the contraction for different times by the following formula:

$$\frac{v_s}{v_p} = \frac{d_c \frac{V_s}{V_p} - d_p}{d_c - d_p}, \tag{III-2}$$

where d_c is the density of the solid metal and d_p is the density of the compact before sintering.

The variation of v_s/v_p with sintering time was obtained by means of the dilatometer for bodies of copper, iron, nickel, cobalt, and silver powders. The samples were compacted to a relatively low density ($0.4-0.5d_c$) so that the densification process would not be disturbed by phenomena inducing expansion.

The overall character of the curves matched the normal course of shrinkage with time, which is well known from experimental work and powder metallurgy practice (reduction of porosity at a high rate in the beginning and abrupt slowdown of the process during sintering). After a series of unsuccessful attempts to derive a very precise mathematical formula directly for the variation of v_s/v_p with τ, the rate of reduction in the volume of pores as a function of the reduction in the volume of pores attained was studied experimentally, i.e., $dv/d\tau = f(v)$. A simple power relationship was found between the rate of reduction in the volume of pores and the reduction in volume of pores attained in a given sintering time.

After sintering time τ_1 the volume of pores was v_1 and the rate of reduction in the volume of pores was $(dv/d\tau)_1$, and at τ_2 these values were, respectively, v_2 and $(dv/d\tau)_2$, so that

$$\frac{\left(\frac{dv}{d\tau}\right)_1}{\left(\frac{dv}{d\tau}\right)_2} = \left(\frac{v_1}{v_2}\right)^n. \tag{III-3}$$

In other words, the ratio of the rates of reduction in the volume of pores proved to be equal to the n-th power of the ratio of the volumes of pores.

If the rate of the specific reduction in the volume of pores at the beginning of isothermal sintering is introduced into this equation then the rate of the specific reduction in the volume of pores after the volume has decreased to v will be equal to

$$\frac{dv}{d\tau \cdot v} = -q\left(\frac{v}{v_{in}}\right)^m. \tag{III-4}$$

where v_{in} is the volume of pores and $q = dv/d\tau \cdot v_{in}$ is the rate of the specific reduction in the volume of pores at the time the isothermal process begins, with $\tau = 0$ and $m = n - 1$.

Relationship (III-4) was found in plotting the variation of $dv/d\tau \cdot v$ with v/v_{in} in logarithmic coordinates: In all cases the experimental points formed a more or less straight line. A diagram for copper powder is shown as an example in Fig. 11.

It was later found that in the absence of effects disrupting the densification process the relationship expressed in (III-4) holds for single-phase bodies under practically any sintering conditions. The equation holds both for materials showing a rapid change in the rate of densification and for powders where the densification rate decreases relatively slowly. The rate at which the densification rate decreases depends on the value of m.

Integration of Eq. (III-4) leads to an equation describing the change in the volume of pores with time

$$v = v_{in}(qm\tau + 1)^{-\frac{1}{m}}, \tag{III-5}$$

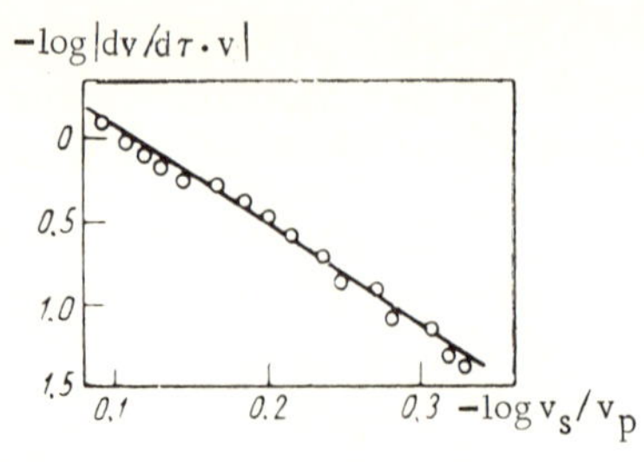

Fig. 11. Variation of $dv/d\tau \cdot v$ with v/v_{in} in logarithmic coordinates (sintering of copper at 800°C).

where v_{in} is the volume of pores at the beginning of isothermal sintering, τ is the isothermal sintering time, q is a coefficient expressing the rate of reduction per cm^3 of the volume of pores at the beginning of isothermal sintering (at $\tau = 0$), and m is a dimensionless constant.

The values of v and v_{in} can be expressed in any units, including relative units. In the calculations below the volume of pores in the compact v_p was used as the unit of measurement. The value of v, determined in this case from Eq. (III-5), was equal to v_s/v_p, i.e., the ratio of the volume of pores after isothermal holding for some time τ, while v_{in} is equal to the same ratio at the beginning of isothermal sintering (at $\tau = 0$).

TABLE 2. Variation of v_s/v_p with τ for Copper Powders

Sintering time, min		Values of v_s/v_p			
		Powder 1*		Powder 2†	
Total	Isothermal sintering time, τ	Exper.	Calc.	Exper.	Calc.
8	0 (c.p.)‡	0.892	0.892	0.716	0 716
22	14	0.850	0.846	0.590	0.587
38	30 (c.p.)	0.833	0.833	0.541	0.541
90	82	0.816	0.816	0.478	0.479
128	120 (c.p.)	0.810	0.810	0.457	0.457
240	232	0.799	0.798	0.418	0.420
360	352	0.787	0.792	0.398	0.398
480	472	0.780	0.787	0.385	0.384

*Obtained by precipitation of zinc from copper sulfate solution and heating in air at 500°C, apparent density 1.72 g/cm³, sintering temperature 765°C, q = 1.15 h^{-1}, m = 48.5

†Obtained by reduction of oxides at 400°C, apparent density 1.87 g/cm³, sintering temperature 825°C, q = 1.93 h^{-1}, m = 7.68.

‡c. p. = calculated point.

TABLE 3. Variation of v_s/v_p with τ for Nickel Powders

Sintering time, min		Values of v_s/v_p			
		Powder 1*		Powder 2†	
Total	Isothermal sintering time, τ	Exper.	Calc.	Exper.	Calc.
8	0 (c. p.)‡	0.671	0.671	0.601	0.601
20	12	0.654	0.656	0.466	0.471
38	30 (c. p.)	0.645	0.645	0.384	0.384
90	82	0.627	0.629	0.286	0.285
128	120 (c. p.)	0.623	0.623	0.251	0.251
180	172	0.616	0.617	0.220	0.222
240	232	0.610	0.611	0.195	0.200
300	292	0.607	0.607	0.177	0.184
360	352	0.606	0.604	0.165	0.172

*Obtained by reduction of oxides in hydrogen at 500°C, apparent density 1.45 g/cm^3, sintering temperature 800°C, q = 0.17 h^{-1}, m = 33.5.
†Obtained by dissociation of nickel carbonyl, apparent density 1.62 g/cm^3, sintering temperature 850°C, q = 1.72 h^{-1}, m = 2.65.
‡c. p. = calculated point.

Thus, let us keep in mind that $v = v_s/v_p$ at time τ and $v_{in} = v_s/v_p$ at $\tau = 0$.

To determine constants q and m from the experimental data we used the method of selected points. In all calculations the selected points were experimentally established coordinates of the points for isothermal sintering times of 0, 30, and 120 min.

The theoretical curves calculated by (III-5) were compared with the experimental data by means of plotting v_s/v_p vs. τ from the dilatometric data for different metal powders at different temperatures. Since the comparison does not give clear results for short periods of time, sintering was conducted for many hours. The changes in the value of v_s/v_p during sintering of different metal powders for 6-8 h are given in Tables 2-6. The experimentally determined values of v_s/v_p for different sintering times and the values of v_s/v_p calculated from Eq. (III-5) are given in the tables. The constants q and m were found from the coordinates of points corresponding to 0, 1/2, and 2 h of sintering, and therefore the fitness of Eq. (III-5) must be judged from comparison of the calculated and actual values of v_s/v_p among these points and beyond their limits. Of particular interest is a comparison at some dis-

TABLE 4. Variation of v_s/v_p with τ for Iron Powders

Sintering time, min: Total	Sintering time, min: Isothermal sintering time, τ	Values of v_s/v_p: Powder 1*, Exper.	Powder 1*, Calc.	Powder 2†, Exper.	Powder 2†, Calc.
8	0(c.p.)‡	0.920	0.920	0.648	0.648
20	12	0.840	0.840	0.531	0.538
38	30(c.p.)	0.798	0.798	0.470	0.470
90	82	0.749	0.749	0.393	0.391
128	120(c. p.)	0.730	0.730	0.363	0.363
180	172	0.719	0.713	0.335	0.338
240	232	0.707	0.698	0.314	0.317
300	292	0.700	0.687	0.300	0.303
360	352	0.692	0.679	0.286	0.291

*Obtained by reduction of scale in hydrogen at 650°C, apparent density 1.96 g/cm³, sintering temperature 700°C, $q = 0.99\ h^{-1}$, m = 14.43.

†Obtained by reduction of oxide in hydrogen at 600°C, apparent density 1.75 g/cm³, sintering temperature 855°C, $q = 1.49\ h^{-1}$,

‡c. p. = calculated point.

TABLE 5. Variation of v_s/v_p with τ for Cobalt Powders

Sintering time, min: Total	Sintering time, min: Isothermal sintering time, τ	Values of v_s/v_p: Powder 1*, Exper.	Powder 1*, Calc.	Powder 2†, Exper.	Powder 2†, Calc.
8	0 (c.p.)‡	0.983	0.983	0.450	0.450
20	12	0.965	0.969	0.359	0.365
38	30 (c.p.)	0.955	0.955	0.327	0.327
90	82	0.934	0.932	0.287	0.286
128	120 (c. p.)	0.922	0.922	0.271	0.271
180	172	0.913	0.912	0.255	0.258
240	232	0.904	0.903	0.241	0.247
300	292	0.895	0.896	0.232	0.239
360	352	0.887	0.891	0.226	0.233

*Obtained by reduction of oxides at 600°C, apparent density 0.88 g/cm³, sintering temperature 765°C, $q = 0.090\ h^{-1}$, m = 27.8.

†Obtained by reduction of oxides at 450°C, apparent density 0.49 g/cm³, sintering temperature 850°C, $q = 2.34\ h^{-1}$, m = 6.92.

‡c. p. = calculated point.

TABLE 6. Variation of v_s/v_p with τ for Silver Powders*

Sintering time, min		Value of v_s/v_p at sintering temperatures of			
		750°C†		800°C‡	
Total	Isothermal sintering time, τ	Exper.	Calc.	Exper.	Calc.
8	0 (c. p.)§	0.693	0.693	0.662	0.662
20	12	0.622	0.613	0.535	0.530
38	30 (c.p.)	0.594	0.594	0.477	0.475
90	82	0.572	0.574	0.415	0.417
128	120 (c.p.)	0.566	0.566	0.394	0.396
180	172	0.560	0.559	0.373	0.377
240	232	0.555	0.553	0.358	0.362
300	292	0.552	0.549	0.351	0.351
360	352	0.548	0.545	0.347	0.342

*Powders obtained by deposition of silver from ammonia solution of copper, apparent density 1.24 g/cm³.
†q = 5.60 h^{-1}, m = 28.49.
‡q = 2.78 h^{-1}, m = 7.24.
§c.p. = calculated point.

tance from the calculated (selected) points. For this purpose, similar experiments with sintering for 50 h were made with nickel, cobalt, and iron powders. The results of these tests are given in Table 7.

In the overwhelming majority of tests there was satisfactory agreement between the theoretical and experimental values of v_s/v_p. Thus, in 90% of the cases the deviation did not exceed ±2.8% after sintering for 6 h or ±5.5% after sintering for 50 h.

The experimental plots of v_s/v_p vs. τ were also used to judge the fitness of empirical equations and those derived from theoretical concepts for describing the densification process. Equation (III-5) was proposed by the present author in 1950 [13]. Before that time densification had been described by mathematical expressions in only a few works. According to Frenkel' [14], a spherical pore in a viscous medium should contract at a constant rate of reduction of the radius $dr/d\tau = -A$. Hence, for a body with spherical pores we have $v = v_{in}(1-B\tau)^3$, where τ is time and B is a constant.

It follows from this equation that after sintering for time $\tau = 1/B$ the pores should disappear. For commonly observed

TABLE 7. Variation of v_s/v_p with τ for Prolonged Sintering

Sintering time, h: Total	Isothermal time, τ	Copper* Exper.	Copper* Calc.	Iron† Exper.	Iron† Calc.	Nickel‡ Exper.	Nickel‡ Calc.	Cobalt§ Exper.	Cobalt§ Calc.
0.13	0 (c. p.)**	0.799	0.799	0.722	0.722	0.706	0.706	0.613	0.613
0.33	0.20	0.703	0.697	0.624	0.631	0.691	0.694	0.516	0.506
0.53	0.50 (c. p.)	0.657	0.657	0.578	0.578	0.683	0.683	0.478	0.478
1.5	1.37	0.611	0.612	0.519	0.516	0.665	0.664	0.441	0.441
2.13	2.00 (c. p.)	0.596	0.596	0.493	0.493	0.656	0.656	0.427	0.427
3	2.87	0.582	0.581	0.464	0.472	0.653	0.648	0.415	0.415
4	3.87	0.569	0.569	0.441	0.454	0.646	0.641	0.407	0.406
6	5.87	0.556	0.551	0.407	0.431	0.635	0.631	0.384	0.391
10	9.87	0.538	0.531	0.374	0.404	0.622	0.618	0.360	0.374
15	14.9	0.522	0.515	0.360	0.384	0.609	0.608	0.346	0.361
20	19.9	0.512	0.504	0.352	0.370	0.600	0.601	0.340	0.353
30	29.9	0.486	0.490	0.345	0.351	0.588	0.591	0.328	0.341
40	39.9	0.477	0.479	0.341	0.338	0.580	0.584	0.320	0.333
50	49.9	0.466	0.472	0.338	0.329	0.572	0.579	0.312	0.327

*Obtained by reduction of oxides in hydrogen at 400°C, apparent density 2.84 g/cm³, sintering temperature 800°C, $q = 1.97\ h^{-1}$, $m = 13.67$.

†Obtained by reduction of oxides in hydrogen at 600°C, apparent density 1.75 g/cm³, sintering temperature 800°C $q = 120\ h^{-1}$, $m = 7.82$.

‡Obtained by reduction of oxides in hydrogen at 550°C, apparent density 1.56 g/cm³, sintering temperature 800°C, $q = 0.102\ h^{-1}$, $m = 24.3$

§Obtained by reduction of oxides in hydrogen at 450°C, apparent density 0.49 g/cm³, sintering temperature 800°C, $q = 3.105\ h^{-1}$, $m = 11.95$.

**c.p. = calculated point.

initial rates of reduction in the volume of pores the calculation gives values of time for complete disappearance of pores from 0.5 to 10 h. The completion of densification within the limits of this (and longer) time does not, in fact, occur. For the case of interconnecting pores it is more valid to use Frenkel's equation for coalescence of spherical particles [14]. In this case the process of coalescence of particles and reduction of the voids between them before the formation of closed pores should occur in a short time (at a high initial rate of the process), which does not occur.

Frenkel's equation can also be used to calculate densification for a model of a ductile body with spherical pores of different sizes. Such a model should undergo densification at a decreasing rate, since the rapidly reduced small pores soon disappear and the rate of subsequent densification will depend on pores of large diameter. However, the actual differences possible in the size of pores (in combination with the total volume of pores of each size) cannot produce the rapid reduction of the densification rate observed in sintering of metal powders. This question will later be examined in more detail (Chapter 10).

The equation proposed by Shaler and Wulff [15] for the reduction in the radius of pores, taking into account the changing density of the medium, $dr/d\tau = -Ad^2$, leads to a complex form of relationship $v = f(\tau)$, analysis of which indicates that pores should close up still more rapidly than follows from Frenkel's equation. This equation is unsuitable for describing the densification of real powders.

On the basis of very general concepts of the proportionality of the rate of decrease in the number of mobile atoms responsible

TABLE 8. $v_s/v_p = f(\tau)$ for Copper Powders Sintered at 800°C, Calculated by Kinetic Equations from Different Authors

Sintering time, h		Value of v_s/v_p					
			Calculated by equation				
Total	Isothermal time, τ	Exper.	(III-5)[1]	(III-7)[2]	(III-9)[3]	(III-11)[4]	(III-17)[5]
0.13	0 (c.p.)*	0.799	0.799	0.799	0.799	0.799	0.718
0.33	0.20	0.703	0.698	0.724	0.687	0.689	0.679
0.53	0.50 (c.p.)*	0.657	0.657	0.657	0.657	0.657	0.657
1.50	1.37	0.611	0.610	0.589	0.612	0.611	0.613
2.13	2.0 (c.p.)*	0.596	0.596	0.596	0.596	0.596	0.596
4	3.87	0.569	0.569	0.591	0.560	0.562	0.548
8	7.87	0.545	0.540	0.586	0.511	0.522	0.476
14	13.87	0.525	0.518	0.583	0.466	0.487	0.397
24	23.87	0.501	0.498	0.583	0.416	0.453	0.297
40	39.87	0.477	0.479	0.583	0.363	0.419	0.174
50	49.87	0.466	0.472	0.583	0.336	0.405	0.110

1) $v = v_{in}(qm\tau + 1) - 1/m$. 2) $v = v_{in} - B[1 - \exp(-K\tau)]$. 3) $v = v_{in}(1 - K\tau^n)$. 4) $v = v_{in}(K\tau^n - 1)^{-1}$. 5) $v = v_{in}(1 - K\tau^{1/2})$.
*c.p. = calculated point.

for shrinkage to the number of these atoms, Bal'shin [2] proposed the following expression for the rate of shrinkage in bulk:

$$\frac{dV}{d\tau \cdot V_0} = -a \cdot \exp(-K\tau). \tag{III-6}$$

In view of the fact that the absolute rate of bulk shrinkage is numerically equal to the rate of reduction in the volume of pores, one can obtain from Eq. (III-6) an expression for the variation in the volume of pores with time:

$$v = v_{in} - B\left[1 - \exp(-K\tau)\right]. \tag{III-7}$$

The constants B and K were determined by the method of selective points, using the coordinates of points corresponding to 0, 1/2, and 2 h of isothermal sintering, as in calculating the constants in Eq. (III-5). The respective calculated values of v/v_p vs. τ are compared with the experimental data in Table 8. The value of v calculated by Eq. (III-7) soon ceases to decrease, while the experimentally observed densification continues not only after 50 h but, as will be shown below, even after 500 h of sintering.

After 1950 many investigators proposed similar relationships in which linear or bulk shrinkage (or the "densification parameter" found from the density or porosity) was a power function of time. These equations are suitable for rough descriptions of the process, although they are not precise enough to describe the entire densification process including rapid reduction in the volume of pores (or volume of the sintered body) at the beginning of isothermal sintering.

In most cases these equations can be reduced to the form of $v/v_{in} = f(\tau)$ by simple mathematical transformation.

Fedorchenko and Andrievskii [16] found that the variation of shrinkage with sintering time is approximated by the equation

$$\frac{\Delta V}{V_{in}} = \frac{\Delta V_0}{V_{in}} + K\tau^n, \tag{III-8}$$

where ΔV is bulk shrinkage, ΔV_0 is the shrinkage at the beginning of isothermal sintering, V_{in} is the initial volume before sintering, τ is the time in which shrinkage increases from ΔV_0 to ΔV, and K and n are constants.

Passing to the change in the volume of pores during isothermal sintering, we obtain

$$v = v_{in}(1 - K' \tau^n), \quad \text{(III-9)}$$

where $K' = K \cdot V_{in}/v_{in}$, v is the volume of pores at time τ, and v_{in} is the volume of pores at the beginning of isothermal sintering with $\tau = 0$.

Determining constants K' and n by means of the selected points and plotting the curve of Eq. (III-9), one is easily convinced of the substantial disagreement between the calculated and experimental values of v_s/v_p after even 3 h of sintering (Table 8). The equation is suitable only for rough calculation of densification in short periods of time, as was noted by the authors of the work cited.

A "densification parameter" was introduced in a kinetic equation by Mäkipirtti [17]:

$$a = \frac{V_p - V_s}{V_p - V_c},$$

where V_p is the volume of the body before sintering, V_s is the volume after sintering, and V_c is the volume of the solid material with no pores.

This parameter is related to the sintering time by an equation of the following form:

$$\frac{a}{1 - a} = (K_0 \tau)^n. \quad \text{(III-10)}$$

Tikkanen [18] confirmed the validity of this equation with sintering of different nickel powders.

Since $V_p - V_s / V_p - V_c = 1 - v_s/v_p$, then, by substituting $K_0^n = K'$ one can obtain Eq. (III-10) in the form

$$v = v_{in}(K' \tau^n + 1)^{-1}. \quad \text{(III-11)}$$

Evidently unaware of the equations cited, Kothari [19] used the completely identical equation

$$\frac{V_p - V_s}{V_s - V_c} = K\tau^n, \quad \text{(III-12)}$$

which is easily reduced to the form of (III-11). Treating test results from sintering tungsten powders differing in particle size, he found a linear relationship in coordinates of log $[(V_p - V_s)/V_s - V_c)]$ vs. log τ, confirming on the whole the validity of (III-12). However, Kothari noted that "the experimental points do not lie exactly on a straight line" and that the observed deviation "indicates the complex nature of the densification process."

Equation (III-11) can be compared with experimental data characterizing the variation of v_s/v_p with τ in sintering for 50 h. Although the experimental and theoretical values almost coincide in the period of the first 4 h, with increasing sintering time these values progressively diverge. The deviation is observed on only one side regardless of the sintering conditions and the material sintered.

The calculated and actual values of v_s/v_p for compacts sintered 50 h (calculations made with the same selected points, corresponding to 0, 1/2, and 2 h of sintering) are given in Table 8. The comparison indicates the more exact description of the densification process by Eq. (III-5).*

Bal'shin proposed a number of rough equations relating the density, porosity, or volume of pores to sintering time [20, 21]. The common basis of these equations is the inverse proportionality of the densification rate (in respective notations) and sintering time. For the rate in the reduction of the volume of pores this relationship has the form

$$\frac{dv}{d\tau \cdot v} = -\frac{1}{m\tau}. \qquad \text{(III-13)}$$

The limits of the applicability of this formula can be found by comparison with a similar expression for $dv/d\tau \cdot v = f(\tau)$ ob-

*Using other methods of finding the constants of the equation, including the method of least squares, one could find values ensuring more even distribution of errors for the points calculated at different sintering times. However, in this case it would be complicated to analyze the suitability of the equation and the reliability of the conclusions would not improve. The method of extrapolating the initial sections of curves to long sintering times is the most suitable for solving the set problem. It should be emphasized again that the same selected points were used to determine the constants of the equations compared.

tained from Eqs. (III-4) and (III-5). Substituting v/v_{in} into (III-4), we obtain its value in terms of τ:

$$\frac{dv}{d\tau \cdot v} = -\frac{q}{qm\tau + 1}. \tag{III-14}$$

If $qm\tau \gg 1$ then, disregarding the units, we obtain the rough equation (III-13). This equation becomes almost exact at large value of product $qm\tau$. The larger qm, the smaller the values of τ for a fairly exact description of $dv/d\tau \cdot v$ vs. τ. For example, for iron powder obtained by reduction of oxides in hydrogen, with $qm = 105\ h^{-1}$, an inversely proportional relationship is observed (with an error less than 5%) after sintering for 12 min, while for nickel powders obtained from nickel carbonyl, with $qm = 4.6\ h^{-1}$, an inverse proportionality (with the same error) is noted only after 4.3 h of sintering. The function $v = f(\tau)$, resulting from (III-13) and having the form

$$v = A\tau^n, \tag{III-15}$$

is suitable in those cases where Eq. (III-13) is suitable. At small values of time the rough equation (III-13) becomes unsuitable. In these cases the rate changes with time by another law, but in conformity with Eq. (III-14). It is very important then that Eq. (III-14) describes the change in the rate of reduction in the volume of pores very accurately both in the beginning of sintering and after many hours of sintering.

Of course, Eq. (III-14), like (III-15), describes densification exactly only in the absence of processes inducing expansion of the body, i.e., under sintering conditions corresponding to constant v_s/v_p at different d_p (see Chapter 1).

We shall not dwell on other kinetic equations of shrinkage [22-29], which are identical to equations (III-3), (III-11), and (III-13), whose suitability has already been discussed.

Among other kinetic relationships let us mention the linear variation of the density with the logarithm of sintering time found by Vasilos and Smith [30] in a study of tungsten powders. This was based on the mathematical formulation by Coble relating the densification rate to the self-diffusion coefficient and grain size before and after sintering [31].

A linear variation of density with the logarithm of sintering time is observed fairly frequently in long-term sintering. However, the limited application of this relationship can be seen from the fact that with $\tau < 1$ the value of the relative density D calculated by

$$D = A + B \ln \tau, \qquad \text{(III-16)}$$

becomes negative. From the resulting time dependence of the densification rate ($dD/d\tau = B/\tau$) it follows that the densification rate at the beginning of isothermal sintering must be infinitely large, which is not justified by experimental or theoretical considerations. Another consequence of this equation – completion of densification after some sintering time at which the value of D reaches unity [according to (III-16), D can increase limitlessly] – is also unconfirmed by experiments. The introduction of τ_0, with no physical meaning, for better agreement of the kinetic equation with the experimental results [30] is hardly justified. This measure only confirms the unsuitability of the equation for $\tau < \tau_0$.

Pines has formulated kinetic laws of shrinkage based on theoretical concepts [32, 33]. According to these concepts, at the beginning of sintering the shrinkage should be proportional to $\tau^{1/2}$ for any powder regardless of sintering conditions. The experimental data indicate that in the beginning of sintering the rate of shrinkage decreases very rapidly with time, and therefore the proportionality of shrinkage with time is hardly observed. The rate of shrinkage can be considered roughly constant only in a very limited time interval. With a long sintering time and $\Delta V/V_0 \sim \tau^{1/2}$ the relationship should be of the following form

$$v = v_{in}\left(1 - A\tau^{\frac{1}{2}}\right). \qquad \text{(III-17)}$$

In logarithmic coordinates $\log[(v_{in} - v)/v_{in}]$ vs. $\log \tau$ this relationship should be expressed as a straight line, the slope of which depends on the exponent $n = 1/2$. The actual value of n in an equation of the type

$$v = v_{in}(1 - A\tau^{n}) \qquad \text{(III-18)}$$

can be determined from the experimental data by means of the relationship

$$n = \frac{\log \frac{v_{in} - v_2}{v_{in} - v_1}}{\log \frac{\tau_2}{\tau_1}} .$$

Using this expression to calculate the value of n in equations for the reduction in the volume of pores during sintering of copper, nickel, and iron powders (see Tables 2-4), we obtain the values of n for different sections of the curves that are given in Table 9.

Thus, the actual value of the exponent in Eq. (III-18) may differ substantially from 1/2. The exponent n does not remain constant for all sections of the curve and the value differs for different powders.

The essential difference in Eqs. (III-5) and (III-18) is that the first indicates the impossibility of attaining complete densification at the end of the time (v approaches zero only with an infinitely large sintering time), while, according to (III-18), complete disappearance of pores is attained after a time equal to $1/A^2$.

Very long isothermal sintering (500 h) has shown that even with such a long sintering time there is still considerable porosity. With interconnected pores the densification slows down but does not cease completely (see Chapter 13). The time for complete disappearance of pores found from values of n and A in Eq. (III-18) calculated for times of 0.5-2 h from the experimental data given in Tables 2-4 amounts to 100-200 h. However, if the value of n is taken as 0.5 (as proposed by Pines) then the time for complete

TABLE 9. Values of n for Different Sections of Densification Curves

Powdered material	Time interval, h	
	0.5—2	20—50
Copper	0.562	0.254
Nickel	0,258	0.162
Iron	0.335	0.060

disappearance of pores is still smaller. The calculation of relationship $v_s/v_p = f(\tau)$ by means of Eq. (III-17) gives a curve beyond the limits of the selected points of the experimental values (see Table 8). It follows that the reduction in the volume of pores during prolonged sintering cannot be described by Eq. (III-17) or the very similar equation (III-18).

Thus, of all the kinetic equations presented above (see Table 8), the reduction in the volume of pores is described most accurately by Eq. (III-5). This equation proved to be suitable for describing the reduction in the volume of pores during sintering with application of outside pressure (hot pressing).

Scholz [34], using the results from [35, 36] and his data on sintering of refractory compounds without application of outside pressure, found that an exponential dependence of porosity on time of the type of Eq. (III-15) is suitable in a wide range of time but not does not hold for the initial section of the densification curve. Relationship $P = f(\tau)$ is more exactly expressed by

$$P = P_{in}\left(1 + \frac{\tau}{\tau_0}\right)^{-K}, \tag{III-19}$$

where τ_0 and K are constants. This expression is similar to (III-5). The author indicates that the physical meaning of constant τ_0 is not clear. Comparison of (III-5) and (III-19) indicates that $\tau_0 = 1/qm$. This relationship does not reveal the physical nature of the constant but explains its origin.

The difference in (III-19) and (III-5) is only that the first describes the change in porosity, i.e., the volume percent of pores in the body, and the second the volume of pores per unit volume of the solid substance. Since v_s/v_p is independent of d_p, preference should be given to (III-5) as descriptive of the process at different d_p.

For the densification of metal and carbide powders during hot pressing Samsonov and Koval'chenko [37] proposed an equation in logarithmic form, which after substitution of porosity for density and some transformations assumes the form of (III-19). The authors indicate that the equation is suitable for describing the first stage of sintering, characterized by a high rate of shrinkage.

The use of (III-19) to calculate the change in the volume of pores during ordinary sintering gives almost complete agreement with the calculations by (III-5) for the first 4-6 h of sintering, and thus the assertion that (III-19) is suitable for sintering combined with application of pressure indicates the accuracy of (III-5) under these sintering conditions. A discrepancy occurs between the curves described by (III-5) and (III-19) only with extrapolation of the initial section of the curve to long sintering times (10 h and more).

It must be assumed that in sintering with application of pressure the formation of closed pores and the elevated gas pressure in these pores are the same as in ordinary sintering and have a substantial effect on the densification process. This is precisely the explanation of the deviation from the curve described by (III-19) during prolonged sintering under pressure [37].

The substantial difference in the course of densification observed by Bal'shin and Trofimova [38] for coarse-grained and fine-grained powders in sintering under pressure is evidently also explained by the substantial effect of the gas pressure in closed pores on the densification process. The closed pores occur during hot pressing of fine powders due to rapid densification after even a few minutes of sintering. In our opinion, the apparent lack of dependence of the densification rate for fine powders on temperature is also explained by the fact that the rate of shrinkage was measured under conditions quite favorable to the development of the opposing process of expansion: The independence of shrinkage from temperature is observed only in sintering of fine-grained powders and only after rapid and considerable densification, occurring in the first minutes of sintering. Under these conditions the opposing effect of gas pressure on shrinkage increases with temperature due to the more intensive evolution of sorbed gases, which under certain conditions can fully compensate the ordinary accelerating effect of temperature on the shrinkage process. It should be noted that in the first minutes of sintering, i.e., in the presence of interconnected pores, the authors observed the usual variation of the shrinkage rate with temperature [38].

During sintering with application of pressure the densification process in the pure form can be observed only in the period where there are interconnected pores, as in ordinary sintering.

A study of shrinkage under the influence of the opposing effect of the gas pressure in closed pores can give kinetic curves quite different in shape, including curves with a minimum corresponding to the replacement of shrinkage with expansion or an undulating curve reflecting the slowdown of the process after the formation of closed pores and later acceleration of the process after reduction of the gas pressure due to diffusion.

An investigation of densification should be made under conditions excluding or holding to a minimum the effect of processes inducing expansion. It should be kept in mind that along with the gas pressure in closed pores, as noted earlier, expansion at the beginning of heating may be induced by elastic-plastic changes in the shape of particles (phenomenon of the elastic-plastic aftereffect during heating of a deformed body). When it is noted, the latter effect must be determined quantitatively and taken into account in determining the shrinkage. With exclusion of the effect of expansion processes the goal of developing general rules of densification becomes simple and quite practicable.

If the processes of densification and expansion always coexisted and developed simultaneously then it would be difficult to explain the possibility of describing the reduction in the volume of pores (including rapid reduction in the beginning and slow reduction at the end of sintering) by means of one simple equation. Here it is appropriate to cite the opinion of Bal'shin that it is impossible to describe the process of densification by one equation [20, 21, 38]. This opinion was due directly to the concept that processes associated with densification and ex-

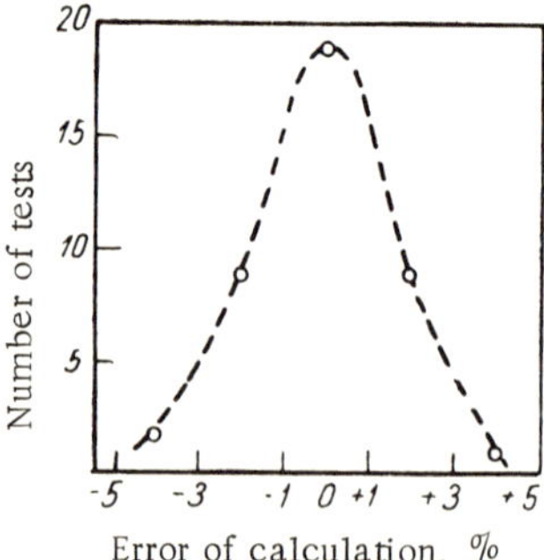

Fig. 12. Distribution of relative error with number of experiments for values of v_s/v_p calculated from Eq. (III-5) in sintering (6 h) of copper, nickel, cobalt, iron, and silver powders (40 tests in all).

pansion coexist under any sintering conditions. At the same time, conditions can be found where the expansion process does not develop to any notable extent and densification occurs in fairly pure form (isothermal sintering in the presence of interconnected pores). Under these conditions the course of densification can ordinarily be described exactly by Eqs. (III-4) and (III-5).

An additional characteristic of the fitness of Eq. (III-5) is the distribution of errors in calculated values of v_s/v_p in sintering for 6 h, shown in Fig. 12. The calculations were made with use of constants q and m found from points for 0, 1/2, and 2 h of isothermal sintering. The errors (difference between the calculated and actual values of v_s/v_p expressed in percent of the value of v_s/v_p) were determined in 40 tests, with plotting of shrinkage curves for copper, nickel, cobalt, iron, and silver powders. The curve of the errors is symmetrical with respect to the experimental value of v_s/v_p taken as 100%, which indicates the random nature of the errors. Also, the maximum error does not exceed 5%, while the standard deviation of the calculated value is only 1.8%.

The suitability of Eq. (III-5) for describing the densification process in very different sintering conditions forces one to the conclusion that there is a connection between the kinetic law expressed by this equation and the densification mechanism. The problem of revealing this relationship has not been fully resolved. However, analysis of Eq. (III-5) leads to important phenomenological generalizations that can serve as the basis for stricter physical hypotheses or theories in the future.

Chapter IV

Phenomenological Importance of the Constants of the Kinetic Equation and Their Dependence on Temperature

Replacing v in Eq. (III-4) with v_{in} (volume of pores at the beginning of isothermal sintering) or $\tau = 0$ in Eq. (III-14), we find that $q = -dv/d\tau \cdot v_{in}$. It follows that constant q is the rate of reduction in the volume of pores at the beginning of isothermal sintering. Its dimension is h^{-1}.

The dimensionless constant m characterizes the rate of the decrease in the rate of reduction in the volume of pores during sintering. If the initial rate of reduction in the volume of pores and the initial volume of pores are known, then further changes in the volume of pores with time will depend on the value of m. Since the specific rate of reduction in the volume of pores at any moment of sintering of a given powder does not depend on the initial density of the compact, the value of m is a constant of the original powder under given sintering conditions (temperature, atmosphere, pressure).

To calculate the constants in Eq. (III-5) it is necessary first of all to make experimental plots of the curve to establish the beginning of isothermal sintering. The selection of a point on the curve corresponding to the beginning of isothermal sintering is almost arbitrary. Ordinarily the beginning of isothermal sintering is selected to ensure that a constant temperature has already been attained.

The value of q depends on the beginning point selected. The later in time from the actual beginning of isothermal sintering that the arbitrary point is selected as the beginning, the smaller the value of q (since the rate of densification decreases rapidly with sintering time). Also, the value of q depends greatly on the rate of increase in temperature up to the beginning of isothermal sintering.

Despite the fact that the dependence of the densification rate on temperature is confirmed by simple tests as well as everyday powder metallurgy practice, it is a far from simple task to explain the temperature dependence of the densification rate at the beginning of isothermal sintering. It is complicated by the fact that substantial changes in the properties of the sintered body (including substantial reduction of the porosity) occur during heating to the isothermal sintering temperature.

It can be considered established that the densification rate during sintering is affected by imperfections in the crystal lattice [32, 39-41]. The extent of imperfections in the lattice changes rapidly at the beginning of heating, the rate of this process also depending on temperature. Thus, the higher the sintering temperature, the more the properties of a crystalline substance change up to the beginning of isothermal sintering. These changes themselves have a considerable effect on the densification rate at the beginning of sintering.

The phenomena occurring in the initial period of sintering will be examined in detail. Here, let us note only that the densification rate at the beginning of isothermal sintering is a very complex function of temperature because of the profound changes occurring in the sintered body in the course of heating to sintering temperature.

In some cases, mainly in sintering of low-activity powders whose properties undergo no substantial changes in the course of heating, one can get a rough idea of the dependence of the densification rate on temperature at the beginning of isothermal sintering. For such powders the value of q increases more or less evenly with temperature. Below, we give values of q and m for copper powders with an apparent density of 2.87 g/cm^3 determined from $v_s/v_p = f(\tau)$ during sintering of compacts under different temperature conditions:

Sintering temperature, °C	750	800	850	900	950	1000	1050
q	0.53	1.15	1.39	1.63	1.47	2.09	2.16
m	18.15	13.74	8.66	6.23	3.60	3.18	2.54

On plots of log q vs. 1/T the experimental points are frequently located along a straight line, indicating an exponential relationship. Hence, one can find the value of E_A, which we call the apparent activation energy of the reduction in the volume of pores. However, it is easily demonstrated that the quantitative characteristic of the temperature dependence is closely connected with the conditions of heating to isothermal sintering temperature. If another time from the beginning of heating is used in the calculations as the beginning of isothermal sintering then the dependence of q on temperature will be determined by the other value of E_A. Thus, it can be assumed that the initial densification rate can be correctly determined only with instantaneous heating to the given temperature. Of course, such heating is not possible in practice. However, several conclusions can be drawn from measurements of the densification rate with very rapid increase of the temperature. Since it is impossible to achieve an instantaneous change to the isothermal sintering temperature by heating at a high rate, the measurement of the densification rate was made with increasing temperature, and the average rate of reduction in the volume of pores was determined within the limits of a given short time interval.

Two series of experiments were set up on sintering with different heating rates (a dilatometric tube was pushed into a furnace heated to different temperatures). From two measurements of the rate in the reduction of the volume of pores in equal time periods but at different temperature intervals we determined the value of the apparent activation energy E_A. The value of E_A increased rapidly with increasing heating rates, i.e., with decreasing time of heating to the temperature range in which the rate of reduction in the volume of pores was determined (Table 10).

It can be assumed that with further increase of the heating rate there would be an additional increase of E_A. Indirect confirmation of this was given by measurements of the densification rate with a rapid increase of the sintering temperature after some iso-

TABLE 10. Effect of Heating Time on Value of Apparent Activation Energy E_A of Reduction in Volume of Pores During Sintering of Copper Powder

Time range from beginning of heating, min	Measurement 1		Measurement 2		E_A, cal/g-atom
	t_{av}, °C*	$dv/d\tau \cdot v$, h^{-1}	t_{av}, °C*	$dv/d\tau \cdot v$, h^{-1}	
8.0—8.5	800	1.15	1000	2.09	8,150
3.0—3.5	916	3.67	1035	6.38	15,460
2.5—3.0	895	3.45	1031	9.00	21,300
2.0—2.5	857	2.31	1000	14.59	36,750

*t_{av} is the average temperature in the range of measurement.

thermal holding (tests with a stepped increase of temperature). In the first stage of isothermal sintering there is a considerable reduction in the rate of shrinkage. The rate at which imperfections are eliminated also decreases sharply. With a rapid increase of the sintering temperature to 150-200°C there is not time for the properties to undergo substantial changes and the effect of temperature on the densification process occurs in pure form. Studies made with copper, nickel, cobalt, and silver powders showed that the temperature dependence of the densification rate with an increase of temperature up to the following temperature stage is determined by the value of the activation energy, which is close in value or even higher than the activation energy of self-diffusion of the given metal. The existing experimental data along with an analysis of these data are given in Chapter 11.

TABLE 11. Values of m in (III-5) for Nickel and Cobalt Powders at Different Sintering Temperatures

Sintering temperature, °C	Nickel			Cobalt reduced from oxides in hydrogen at	
	Reduced from oxides in hydrogen at		Carbonyl		
	600 °C	400 °C		600 °C	450 °C
650	—	—	6.73	—	65.9
700	106.6	73.0	—	—	—
750	87.3	69.3	3.44	17.8	16.6
800	36.2	33.5	—	9.97	11.9
850	15.9	22.0	2.52	5.98	6.92

Under ordinary sintering conditions, however, it is often impossible to discern a regular variation of q with temperature because of the considerable changes in the properties up to the beginning of isothermal sintering. An example of the regular changes in this value with an increase of the isothermal sintering temperature is given below in values of q determined from the variation of v_s/v_p with τ for nickel powder with an apparent density of 0.86 g/cm^3 obtained by reduction of nickel oxide.

Sintering temperature, °C	700	750	800	850	900	950	1000
q	25.24	0.11	0.06	0.13	0.30	1.31	2.44
m	122.30	65.60	23.61	16.18	5.79	4.11	2.98

In contrast to q, the constant m varies distinctly with temperature – it decreases continuously with increasing temperature. This change in the value of m means that the drop in the rate of reduction in volume of pores with time (or the rate of the "slow-down") decreases with increasing temperature. The values of m found in sintering nickel and cobalt powders at different temperatures are given in Table 11.

To determine the variation of m with temperature more precisely, large numbers of plots were made for the densification of many powders at different temperatures. Figure 13 shows a set of curves obtained by means of a dilatometer for copper powders, and Fig. 14 for nickel powders. The values of q and m for each curve are given in the legends. The results for another

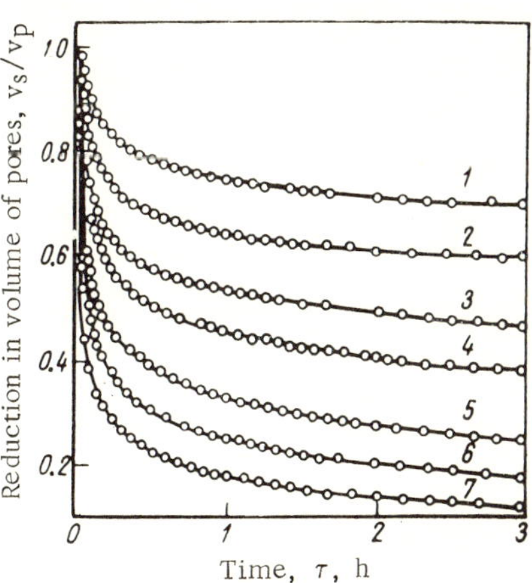

Fig. 13. Variation of v_s/v_p with τ for copper powder with an apparent density of 2.87 g/cm^3 at different sintering temperatures. 1) 750°C, q = 0.53, m = 18.15; 2) 800°C, q = 1.15, m = 13.74; 3) 850°C, q = 1.39, m = 8.66; 4) 900°C, q = 1.63, m = 6.23; 5) 950°C, q = 1.47, m = 3.60; 6) 1000°C, q = 2.09, m = 3.18; 7) 1050°C, q = 2.16, m = 2.54.

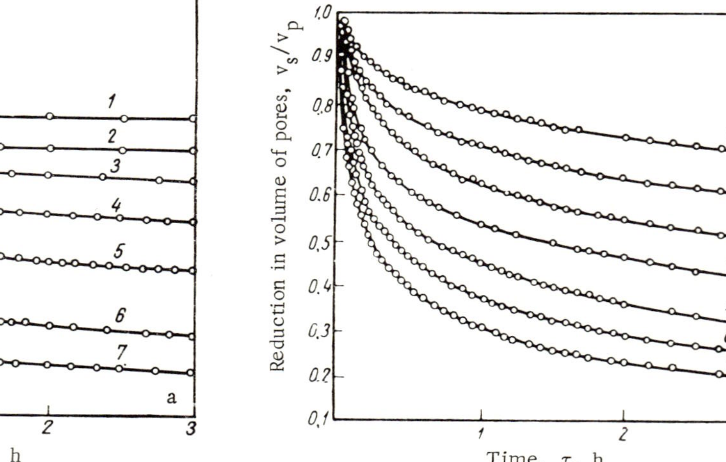

Fig. 14. Variation of v_s/v_p with τ for nickel powders at different sintering temperatures. a) Powder reduced from oxides, apparent density 0.86 g/cm^3: 1) 700°C, q = 25.24, m = 122.30; 2) 750°C, q = 0.11, m = 65.60; 3) 800°C, q = 0.06, m = 23.61; 4) 850°C, q = 0.13, m = 16.18; 5) 900°C, q = 0.30, m = 5.79; 6) 950°C, q = 1.31, m = 4.11; 7) 1000°C, q = 2.44, m = 2.98. b) Carbonyl powder, apparent density 1.63 g/cm^3: 1) 700°C, q = 0.39, m = 7.62; 2) 750°C, q = 0.52, m = 5.84; 3) 800°C, q = 0.58, m = 4.87; 4) 850°C, q = 1.09, m = 4.43; 5) 900°C, q = 1.24, m = 3.02; 6) 950°C, q = 1.48, m = 2.57; 7) 1000°C, q = 1.64, m = 2.08.

TABLE 12. Calculated and Actual Values of v_s/v_p in Sintering of Copper Powders*

Sintering time, h		Value of v_s/v_p at sintering temperature (℃)															
		685 (q=1.28 h^{-1}, m=19.63)		745 (q=2.70 h^{-1}, m=14.76)		795 (q=5.33 h^{-1}, m=12.80)		850 (q=12.22 h^{-1}, m=5.87)		900 (q=1.07 h^{-1}, m=4.15)		945 (q=1.71 h^{-1}, m=3.31)		1000 (q=1.86 h^{-1}, m=3.01)		1045 (q=1.75 h^{-1}, m=2.21)	
Total	Isothermal, τ	Exper.	Calc.	Exper.	Calc.	Exper.	Calc.	Exper.	Calc.	Exper.	Calc.	Exper.	Calc.	Exper.	Calc.	Exper.	Calc.
8	0	0.934	0.934	0.859	0.859	0.790	0.790	0.679	0.679	0.616	0.616	0.526	0.525	0.441	0.441	0.347	0.347
20	(c. p.)† 12	0.896	0.895	0.795	0.790	0.710	0.703	0.572	0.575	0.521	0.529	0.417	0.418	0.341	0.344	0.261	0.268
38	30	0.867	0.867	0.752	0.752	0.662	0.662	0.514	0.514	0.465	0.465	0.350	0.350	0.283	0.283	0.213	0.213
90	(c. p.)† 82	0.829	0.831	0.704	0.706	0.615	0.615	0.442	0.443	0.386	0.385	0.274	0.273	0.215	0.215	0.152	0.151
128	120	0.816	0.816	0.690	0.690	0.598	0.598	0.417	0.417	0.355	0.355	0.246	0.246	0.192	0.192	0.130	0.130
180	(c. p.)† 172	0.801	0.802	0.673	0.674	0.584	0.582	0.394	0.393	0.332	0.328	0.225	0.222	0.168	0.172	0.111	0.112
240	232	0.792	0.791	0.660	0.660	0.572	0.569	0.376	0.375	0.310	0.307	0.204	0.204	0.152	0.157	0.099	0.099
300	292	0.784	0.782	0.652	0.651	0.562	0.558	0.362	0.360	0.295	0.291	0.186	0.191	0.141	0.145	0.092	0.090
360	352	0.778	0.774	0.646	0.643	0.556	0.550	0.354	0.349	0.286	0.279	0.176	0.181	0.131	0.137	0.085	0.083

*Powder obtained by substitution of zinc for copper in a copper sulfate solution with subsequent reduction in hydrogen at 400°C: apparent density 2.87 g/cm^3.

†c.p. = calculated point.

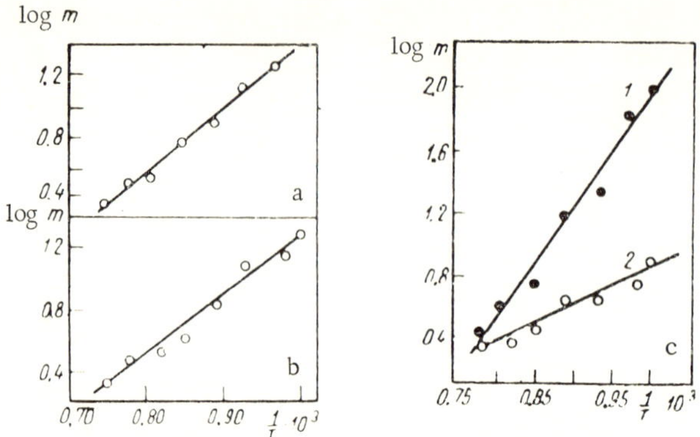

Fig. 15. Variation of log m with 1/T for different powders. a) Copper powder obtained by reduction of oxides in hydrogen at 500°C, apparent density 2.87 g/cm^3; b) copper powder obtained by replacing copper with zinc in copper sulfate solution and additional reduction in hydrogen at 400°C, apparent density 2.35 g/cm^3; c) nickel powder obtained by reduction of oxides in hydrogen at 450°C, with apparent density 0.86 g/cm^3 (1), and obtained by dissociation of carbonyl, with apparent density 1.63 g/cm^3 (2).

series of tests with copper powders prepared under other conditions are given in Table 12. The graphs of log m vs. 1/T plotted from these data (Fig. 15) show a linear variation, which indicates an exponential variation of m with temperature:

$$m = \beta \cdot \exp \frac{U}{RT}. \qquad \text{(IV-1)}$$

We shall not give a more detailed analysis of this relationship (the meaning of U will be discussed in Chapter XIII). Let us note only that a relationship of this type is found without exception in all cases of sintering metal powders in the absence of effects from factors disrupting the normal course of densification.

The existence of a clear relationship between m and the temperature is obviously due to the fact that the value of m is not noticeably dependent on the conditions of heating to the isothermal sintering temperature. The selection of the point corresponding to the beginning of isothermal sintering (as in the case of the other two points selected on the experimental curve) has no effect on the results of calculating m or, consequently, the position of the curve calculated by Eq. (III-5) on the plot of v_s/v_p vs. τ. It is only necessary that the point whose time coordinate is taken as equal to zero lie on the isothermal curve. In this case it can be as far as desired from the actual beginning of isothermal sintering (locating it away from the actual beginning is undesirable only because the change in the volume of pores within the limits of the calculated curve decreases sharply, since the densification rate is highest at the beginning of isothermal sintering).

As follows from Table 11, the value of m may vary within very wide limits. Powders prepared by different methods (carbonyl nickel and powder obtained by reduction of nickel oxide in hydrogen, for example) have greatly different values of m under the same sintering conditions. The temperature dependence of m also differs substantially for such powders: The slope of the lines on graphs of log m vs. 1/T differs greatly (see Fig. 15b), which indicates substantially different values of U in Eq. (IV-1). However, calcination of metal powders before sintering greatly reduces the value of q (determined in rapid heating) but changes the value of m only slightly.

In comparing the changes in q and m it is convenient in some cases to use the value of the initial rate of reduction in the volume of pores in terms of the initial volume of pores, taken as unity. The rate of the reduction in the volume of pores can be extrapolated to the initial volume of pores in the compact by means of Eq. (III-4) from previously found values of q and m:

$$q_0 = \left(\frac{dv}{d\tau \cdot v}\right)_{v=1} = q v_{in}^{-m}, \qquad \text{(IV-2)}$$

where v_{in} is the specific volume of pores (i.e., the value of v_s/v_p) at the beginning of isothermal sintering, q_0 is an arbitrary value (the value of the rate of reduction in the volume of pores at the beginning of sintering in the hypothetical case of instantaneous heating to constant temperature). The calculation assumes that the same changes in the properties of the sintered body will occur in the beginning of sintering at high temperature after instantaneous heating to sintering temperature as in the period of gradual heating, which is obviously untrue. Nonetheless, the calculated value of q_0 is useful for analyzing the densification of different pow-

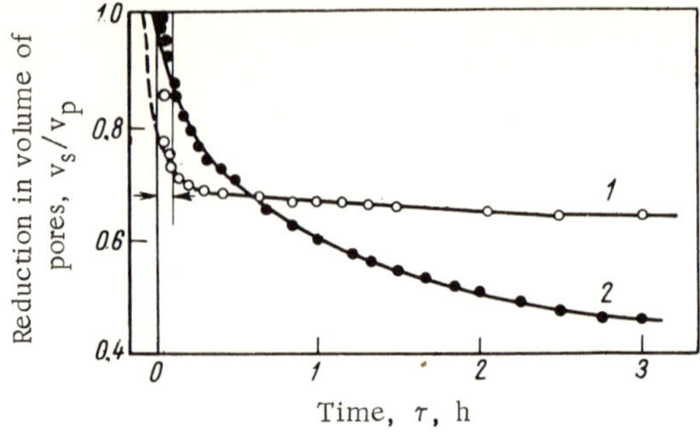

Fig. 16. Variation of v_s/v_p with τ for sintering (800°C) two different nickel powders. 1) Powder obtained by reduction of oxides in hydrogen at 700°C, apparent density 1.68 g/cm^3, q_0 = 47.6, m = 24.3; 2) powder obtained by dissociation of carbonyl, apparent density 1.75 g/cm^3, q_0 = 1.45, m = 3.75. Dashed line obtained by extrapolation of isothermal curves to value v_s/v_p = 1 by Eq. (III-5). The arrows indicate the period of increase in temperature.

ders. Comparison of q_0 and m for different powders makes it possible to resolve the question of whether the same degree of densification can be obtained for these powders under given sintering conditions. The isothermal densification curves may intersect for sintering of compacts prepared from powders produced by different methods. This happens in those cases where the inverse relationship of q_0 and m that is common for varying conditions of sintering is disrupted, i.e., where both values of q_0 and m for one powder are larger than the corresponding values of q_0 and m for another powder. For example, this is the case for sintering of carbonyl nickel powder and powder reduced from nickel oxide (Fig. 16). The curves of v_s/v_p vs. τ intersect after 32 min at 800°C.

Comparison of the curves of v_s/v_p vs. τ for different powders shows that a high initial rate of densification does not necessarily lead to substantial densification at the end of sintering. A high initial rate may slow down later, and the density obtained may be lower than in the case of a low initial rate of densification followed by a smaller drop of the initial rate in the course of sintering.

These observations indicate that the course of densification is not determined by any single parameter of the compact or the original powder (dispersity, specific surface, or integral activity factor).

We shall come back to a phenomenological interpretation of the values of q and m after clarifying a number of characteristics of the densification of metal powders during sintering.

Chapter V

Basic Differences in the Densification Process in Crystalline and Amorphous Bodies

The general rule observed in sintering of metal powders – the proportionality of the rate of reduction in the volume of pores to the mth degree of reduction in the volume of pores attained – obviously reflects several characteristic features of the densification mechanism. Additional information on the characteristics of densification in crystalline powders may permit comparison of the laws governing densification of amorphous and crystalline bodies.

Studies of models of sintered bodies (spherical particles touching each other) indicate that for amorphous and crystalline bodies the coalescence of the particles obeys substantially different laws. For amorphous bodies the increase of the contact surface between spherical particles depends on the relationship $\chi^2 = a\tau$, where χ is the radius of the contact surface; a is the proportionality coefficient; τ is time [24, 42, 43]. This relationship matches Frenkel's equation describing the sintering of ideally viscous bodies [14]. Similar experiments with crystalline bodies (mainly with spherical particles of metals) have established another time dependence. It is most often found that $\chi^5 = a\tau$ [24, 44-46]. In the opinion of many investigators, this relationship indicates that the basic mechanism of the transfer of substance in sintering (coalescence) of spherical particles is diffusion flow, the mechanism of which was examined in detail by Pines [47, 48] and Kuchinskii [44, 49]. Another relationship was found in some investigations: $\chi^7 = a\tau$ (dispersed copper particles at relatively low temperatures [44], spherical particles of iron [50]). According to Kuchinskii, this relationship corresponds to the mechanism of surface self-diffusion [44]. Calculations by other authors, however, do not agree with Kuchinskii's conclusions [48, 51].

It is obviously complicated and not always possible to establish the mechanism of a process from the form of relationship $\chi = f(\tau)$, since theoretical calculations are not undisputed and experimental data for fine particles are rather inaccurate, as was indicated by Thümmler [52]. From the accumulated experimental data it follows only that there is a radical and clear difference in the kinetics of the growth of necks during coalescence of amorphous and crystalline (metal) spheres. In the opinion of many authors, the main process in the growth of the contact is bulk diffusion flow. However, it would be premature to extend this conclusion to sintering of metal powders. The kinetic law for the ensemble of particles resulting from the diffusion mechanism does not correspond to the course of densification in real powders. The mathematical formulation of the time dependence of densification for an ensemble of particles obtained by Kingery and Berg [24] requires that bulk shrinkage be proportional to $\tau^{4/5}$. An exponent close to unity indicates the small change in the densification rate with time. The variation of v with time in this case will have the form

$$v = v_{in}\left(1 - A\tau^{\frac{4}{5}}\right). \qquad \text{(V-1)}$$

From (V-1) it follows that the disappearance of pores or the change from interconnected to closed pores must occur after time of order 1/A (more precisely, $\tau_{fi} = A^{-5/4}$). In the equation for densification of an ensemble of particles of irregular shape derived by Johnson and Cutler [53] this conclusion remains valid, since the exponent remains close to unity.

The absence of any rapid deceleration or rapid completion of densification, resulting from Eq. (V-1), does not correspond to the actual course of densification of crystalline powders, for which a sharp decline in the densification rate is characteristic at the beginning of sintering.

The differences in the kinetics of densification of amorphous bodies and bodies of metal powders are not exhausted by characteristics observed in sintering spherical particles. The essential characteristic of sintering metal powders, which has not been found in experiments with models, is the very high initial densification rate with rapid heating to sintering temperature and the sharp decline of the process while interconnected pores still exist. The probable reason for the high initial densification rate is the influence of the disruption of the normal structure of the crystal lattice (imperfections or defects of the lattice) on the flow rate of the crystalline substance. Before examining the phenomenological manifestations of this relationship we must determine whether it is possible to explain the observed densification pattern of crystalline bodies without these concepts, which are based on the laws governing viscous flow when applied to a porous body with pores of different size, including very fine pores.

Since, according to Frenkel' [14], the rate of reduction in the volume of pores should increase with decreasing volume of pores (from $dr/d\tau = -A$ it follows that $dv/d\tau \cdot v = -Bv^{-1/3}$), there is a basis for the assumption that the high rate of reduction in the total volume of pores at the beginning of sintering is due to the presence of very fine pores in the body, which rapidly disappear and thus affect the densification rate only at the very beginning of sintering. This concept, however, is easily checked by simple calculations employing the densification rate measured in the beginning and at the end of prolonged sintering. Thus, after nickel is sintered 50 h at 800°C the specific rate of reduction in volume of pores is 0.0008 h^{-1} (calculated from experimental data given in Table 7). With rapid increase of temperature at the beginning of sintering, however, the rate of reduction in the volume of pores amounts to around 20 h^{-1} (calculation from data shown in Fig. 17).

According to [14], $dv/d\tau \cdot v = -B/r$ for a spherical pore, where r is the pore radius. Hence it follows that the change in the radius or diameter of the pores must be inversely proportional to the rate of reduction in volume of pores, which determines the densification rate at a given sintering time. After sintering for 50 h the average pore size is of order 10 μ (the size of most pores visible in the microsection is 2-20 μ). This pore size is matched by a rate of reduction in pore volume of 0.0008 h^{-1}. Since the volume of pores at the end of sintering is approximately half the initial volume of pores, the volume of fine pores disappearing in the process of sintering and causing rapid initial densification could

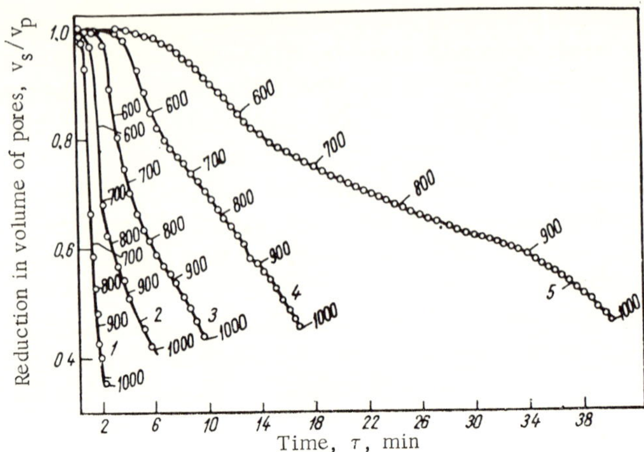

Fig. 17. Variation of v_s/v_p with time for sintering of nickel powder (apparent density 0.86 g/cm^3) obtained by reduction of oxides in hydrogen at 450°C, with heating to sintering temperature (1000°C) at different rates. 1) Sintering temperature reached in 2.0 min; 2) 5.6 min; 3) 9.8 min; 4) 16.5 min; 5) 39.6 min. The temperatures reached in the course of heating are shown on the curves.

not be more than half the volume of pores in the body before sintering. Hence it follows that if the only mechanism of densification were viscous flow then the observed initial rate of reduction in volume of pores (20 h^{-1}) could occur only with the size of fine pores of order $10 \times 0.0008/2 \times 20 \approx 0.0002\,\mu$, or around 2 Å. The inaccuracy of the supposition that large numbers of pores exist that are far smaller than the unit cell of the crystal will be evident. In passing, let us note that similar calculations for cylindrical (but not spherical) pores to explain the high rate of densification at the beginning of sintering require the existence of still more fine pores in the sintered body.

The high initial densification rate cannot be explained without assuming that imperfections in the crystal lattice (defective sections) affect the flow rate. The possibility that imperfections of the crystal structure affect the rate of shrinkage was noted long ago [47], although the kinetic laws governing shrinkage in the first investigations were grounded and mathematically developed without considering the effect of lattice imperfections [14, 47].

In 1950 the present author indicated the impossibility of explaining the kinetic features of the densification process in crystalline substances without taking into account the effect of structural defects on the flow rate of the crystalline substance [39]. This point of view was widely held at a later time. At the present time the substantial effect of the condition of the crystal lattice (concentrations of imperfections or defects) on the rate of reduction in volume of pores or on the rate of volume shrinkage is taken into account in some form or other in almost all work on the kinetics of densification [4, 32, 40, 41, 54].

Leaving aside, for the present, the physical essence of processes determining the relationship, let us consider the phenomena observed in sintering of powders in which the densification rate is affected by the concentration of imperfections.

It should be said first that the overall course of densification of metal powders depends mainly on the imperfection in the crystal and their elimination during heating of the sintered body. In the beginning of sintering a large concentration of imperfections causes a high rate of densification, which decreases rapidly due to simultaneous elimination of the imperfections. During prolonged sintering the concentration of imperfections decreases so much that, despite the considerable interconnected porosity, the densification rate can drop to a very small value. Technologists are quite familiar with the fact that the effect of preliminary calcination of metal powders on the densification rate of compacts formed from calcined powders also reflects the influence of the concentration of imperfections in crystalline particles of the powder on the densification rate (the densification rate decreases so much more than would be expected from slight coarsening of the powder during calcination).

Phenomena in which densification was observed to vary with the concentration of imperfections have been described many times [55, 56]. Below we shall consider mainly those experimental data in which there is a difference in the kinetics of the processes of reduction in volume of pores and elimination of imperfections in the crystal structure. If the first process is directly related to the second, then elimination of imperfections ("healing" or correction of defective sections) may develop independently of the first process. The difference consists first of a difference in the temperature dependence of the processes. Analysis of the experimental data leads to the conclusion that the activation energy of the elimination of imperfections is smaller than the activation energy of flow. The inequality of the activation energies of these processes is always apparent in comparing the course of densification at different temperatures. With an increase of temperature the ratio of the rates of the processes changes substantially: The rate of reduction in volume of pores increases more rapidly than the rate of elimination of impurities with increasing temperature. At low temperatures it is the reverse: The elimination of imperfections may occur at a substantial rate, while the reduction in volume of pores develops slowly. The change in the relationship of the rates of these processes is due to several characteristic features of the densification of crystalline powders at different temperatures.

A typical phenomenon in which the temperature dependence of the elementary processes differs (the flow rates at a constant concentration of imperfections and the rates of disappearance of imperfections) was the previously mentioned (Chapter IV) variation of the densification rate at the beginning of isothermal sintering with the rate of heating to the given sintering temperature.

The experimental data reflecting the change in the ratio of the rates of these elementary processes were also obtained in experiments to determine the course of shrinkage in the process of heating at different rates. Experiments on copper compacts indicated that the densification rate is almost proportional to the rate of increase in temperature. Thus, for slow heating of the compact (in 54.5 min) to 1000°C and with holding in the range of 800-1000°C for 14.8 min the change in the ratio of v_s/v_p is 0.441 for this period (from 0.836 to 0.495). With rapid heating to 1000°C (in 5.5 min) with holding at 800-1000°C for only 2.8 min the same change in the ratio of v_s/v_p was attained, namely 0.450 (in the range of 800-1000°C, v_s/v_p decreased from 0.905 to 0.550). Despite the reduction of the holding time in the range of 800-1000°C by a factor of five, the change in v_s/v_p was approximately the same, which indicates that the rate of reduction in volume of pores increased approximately in proportion to the increase in heating rate. The effect of the time factor was almost unnoticeable in this experiment.

TABLE 13. Effect of Rate of Increase in Temperature on the Reduction in Volume of Pores Attained (v_s/v_p) and the Rate of Reduction in Volume of Pores ($dv/d\tau \cdot v$) for Nickel Powders*

Average rate of increase in temperature, deg/min	Time of heating to 1000°C, min	$dv/d\tau \cdot v$, min^{-1}†	v_s/v_p‡
25	39.8	0.013	0.472
59	16.7	0.031	0.458
100	9.8	0.043	0.438
178	5.5	0.075	0.422
490	2.0	0.265	0.361

*Heated at almost constant rate to 1000°C and then rapidly cooled.
†At 900°C.
‡At 1000°C.

In sintering of nickel powder one also observes acceleration of densification with an increase of the heating rate. The degree of densification reached changed little. It even increased somewhat (v_s/v_p decreased) with reduction of the heating time (see Table 13 and Fig. 17). Although the holding time in the range of 800–1000°C was only 0.4 min in rapid heating, and 15.7 min in slow heating, the densification was somewhat larger in the first case.

Thus, in these experiments the heating rate had a stronger effect on the final densification than the holding time in the temperature range at which most of the densification occurred.

From the curves in Fig. 17 it follows that at some temperature (900°C, for example) the rate of reduction in volume of pores depends on the rate of increase in temperature in the preceding time period (see Table 13). The holding time at low temperature has an inhibiting effect on the densification process during subsequent heating to high temperature. This effect must be due to the elimination of some portion of the imperfections in the absence of substantial densification. With rapid heating and a short holding time in the "unfavorable" temperature range, where imperfections are eliminated with little densification, a substantial concentration of imperfections is retained until a high temperature is reached, which favors a high densification rate.

With slow heating a substantial portion of the imperfections disappear in the low-temperature range, and when high temperatures are reached the rate of shrinkage is low because of the reduced concentration of imperfections.

The presence of an unfavorable temperature range in which the elimination of imperfections occurs at a notable rate with little densification can be established for active powders of any metals as well as by more direct means. For this purpose it is sufficient to compare the densification attained in samples presintered at different temperatures with the densification after final sintering at a higher temperature.

Table 14 shows the reduction in volume of pores due to double sintering of copper compacts. The presintering temperatures were different and the final sintering temperature was 820°C. With an increase of presintering temperature the reduction in volume of pores during final sintering decreased regularly (mainly as the result of the elimination of part of the imperfections during presintering). With presintering in the temperature range of 390-660°C the volume of pores after final sintering was larger (and densification smaller) than after single sintering at 820°C (the total sintering time at high temperature was the same in both cases and equal to 30 min). Presintering at 505°C proved to be the most unfavorable for densification. In this case the volume of pores after final sintering was 20% larger than after single sintering.

TABLE 14. Reduction in Volume of Pores After Single and Double Sintering of Copper Powder Compacts

Presintering temperature, °C	Specific reduction in volume of pores		
	After presintering, $(v_s/v_p)_1$	After sintering at high temperature,* $(v_s/v_p)_2$	Total, $(v_s/v_p)_{1+2}$
—	—	0.340†	0.340†
390	0.992	0.371	0.368
505	0.917	0.469	0.430
610	0.773	0.491	0.380
660	0.625	0.601	0.376
710	0.509	0.639	0.322
785	0.350	0.766	0.268

*820°C, 30 min.
†Single sintering.

TABLE 15. Reduction in Volume of Pores After Single and Double Sintering of Nickel, Cobalt, and Tungsten Carbide

Powder	Sintering method	Sintering temperature, °C		v_s/v_p
		Presintering	Final sintering	
Nickel	Single	—	800	0.466
	Double	700	800	0.553
Cobalt	Single	—	845	0.221
	Double	710	845	0.305
Tungsten carbide	Single	—	1600	0.385
	Double	1400	1600	0.411

Data on the densification of different powders after single or double sintering, with presintering in the unfavorable temperature range, are given in Table 15. It is interesting to note that the total densification is also smaller for nonmetallic crystalline powders (tungsten carbide) when presintering is used. Many other examples could be cited, since this rule is general for crystalline bodies with a nonequilibrium lattice that are capable of densification during sintering.*

It does not follow that any presintering must result in less densification as compared with single sintering. If the presintering temperature is so low that the concentration of imperfections remains almost unchanged or is so high that it approaches the final sintering temperature and substantial densification thus occurs in presintering then the total densification after final sintering may be equal to or larger than the densification in single sintering. A lower total densification after double sintering as compared with single sintering at high temperature is observed, but only in a specific temperature range of presintering for each powder, and depending also on the final sintering conditions.

*It may be noted that not all crystalline bodies are capable of densification during sintering. Thus, porous bodies of several salts with ionic bonds, NaCl in particular, are strengthened during sintering, although the volume is hardly reduced even at temperatures near the melting point [24].

This phenomenon occurs only in sintering of crystalline bodies. It does not occur in sintering of amorphous bodies. A study of the effect of heating rate on densification of amorphous bodies and also the comparison of densification in double and single sintering indicates that the densification rate of amorphous bodies is a simple function of temperature and time. The densification attainable increases with the temperature and time. Thus, with slow heating to a given temperature the densification attained is always larger than with rapid heating. For example, in sintering of powdered glass the densification increases regularly (the value of v_s/v_p decreases) with decreasing heating rates.

The results of measuring v_s/v_p in sintering glass powder with continuous increase of the temperature to 650°C and abrupt cessation of heating are given below:

Heating time, min. .	3	5.5	13	29
v_s/v_p	0.859	0.810	0.747	0.613

Similar results were obtained in heating to 700°C:

Heating time, min .	5	8	12	32
v_s/v_p	0.453	0.419	0.257	0.177

The change in the densification rate of an amorphous body with a change of sintering temperature depends only on the temperature dependence of the flow of the amorphous substance. Therefore, with repeated sintering in any temperature range the reduction in volume of pores attained is the sum of the two, and the total densification after sintering twice is always larger than the densification after single sintering. This is observed in sintering of any amorphous body such as a compact of glass powder, rosin, pitch, etc. The results on the reduction in volume of pores after sintering of glass powder under different conditions are given in Tables 16 and 17. The experiments showed that presintering has almost no effect on the densification during final sintering, while the total densification after final sintering was larger in all cases than after single sintering.

Thus, for amorphous bodies there is no unfavorable temperature range. Prolonged sintering at low temperature is fully equivalent to short sintering at high temperature. Therefore presintering always increases the total densification.

TABLE 16. Reduction in Volume of Pores After Single and Double Sintering of Glass Powder*

Particle size, μ	Sintering method	Specific reduction in volume of pores		
		After pre-sintering, $(v_s/v_p)_1$	After final sintering, $(v_s/v_p)_2$	Total, $(v_s/v_p)_{1+2}$
10—80	Single	—	0.505	0.505
	Double	0.915	0.508	0.465
80—120	Single	—	0.769	0.769
	Double	0.983	0.750	0.736

*Presintering at 590°C, final sintering at 630°C, sintered 30 min each time.

All the phenomenological characteristics of sintering amorphous bodies, in full agreement with studies of models, indicate that densification of amorphous bodies is based on a single elementary process – viscous flow of the substance.

The main phenomenological difference in sintering of porous crystalline and amorphous bodies is the fact that prolonged sintering of crystalline bodies at low temperature is not equivalent to short tempering at high temperature. This difference is due to

TABLE 17. Reduction in Volume of Pores in Subsequent Sintering of Glass Powder with Particle Size of 18-80 μ

Sintering method	Relative reduction in volume of pores		
	After pre-sintering, $(v_s/v_p)_1$	After high-temperature sintering,* $(v_s/v_p)_2$	Total, $(v_s/v_p)_{1+2}$
Single (high-temperature)*	—	0.485	0.485
Double†	0.505	0.545	0.275
Triple‡	0.465	0.506	0.234

*640°C for 30 min.

†Presintering at 630°C for 30 min.

‡Presintering in two stages for 30 min at 590 and 630°C.

the change in the flow rate of crystalline bodies with changes in the concentration of imperfections and the difference in the temperature dependence of the two basic elementary processes – bulk flow of the substance with a constant concentration of imperfections and the elimination of imperfections. It follows that densification of a crystalline body cannot be regarded as a single elementary process, the kinetics of which is determined by one value of the activation energy. The course of densification is a complex function of a series (mainly two) of elementary processes with different activation energies.

It is probable that this is responsible for the ability of crystalline bodies to undergo substantial but not complete densification in an exceedingly wide temperature range. Active metal powders obtained at low temperatures are capable of considerable densification at temperatures of $0.3-0.4T_m$, and sometimes at lower temperature. In addition, an increase of the sintering temperature by 100-200°C (and sometimes more) does not lead to disappearance of interconnected pores. The densification rate decreases long before closed pores occur.

Sintering of amorphous bodies occurs in a narrow temperature range (less than 100°C for glass). With even a small increase of temperature (sometimes 10-20°C) the process ceases to lag in the stage where interconnected pores exist and the body is rapidly densified to a constant density corresponding to complete division of interconnected pores into closed pores. The sintering temperature of amorphous bodies is close to the softening point, where the viscosity of the amorphous substance begins to decrease rapidly.

The possibility of sintering crystalline bodies at such comparatively low temperatures is unquestionably connected with the effect of imperfections on the densification rate, while the common lag of densification in the stage of interconnected pores is due to the decrease of flow because of the rapid reduction in the concentration of imperfections.

Chapter VI

Change in the Surface and Volume of Pores under Various Sintering Conditions

In the preceding chapter we examined the phenomenological manifestations of the two elementary processes – bulk flow and elimination of imperfections in crystals – that are primarily responsible for the course of densification. During sintering there are signs of the development of still another elementary process – migration of the substance along the surface of pores (self-diffusion surface). The outward manifestation of this process is a substantial increase in the gas permeability of porous bodies after sintering at low temperature, which is due to the smoothing out of the surface relief and the simplification of the shape of interconnected pores.

Surface migration of the substance leads to simplification of the shape of pores and reduction of the surface of pores, although it has no effect on the volume of pores [24, 39]. Let us recall the basic assumptions bearing on the absence of any direct influence of surface migration of the substance on changes in the volume of pores:

1. From purely geometric considerations, the transfer of material from one section of the surface to another cannot change the volume of pores.
2. Migration of the substance along the surface of particles, filling in of gaps at contacts, and the growth of contacts by this means do not cause the centers of particles to draw closer together. In order for the particles to be bound together it is necessary that the material be re-

> moved from zones between the centers of adjacent particles, which is possible only with bulk flow of the material.

At the same time, it should be kept in mind that bulk flow also affects the relief of pore surfaces. The difference in capillary pressures around sections of the surface with different curvatures (protrusions and depressions) ensures filling of the depressions and leveling of the protrusions due to bulk flow of the material. Thus, the surface relief of pores can be smoothed out by two elementary processes – surface migration and bulk flow of the material – but a change in the volume of pores results only from bulk of the material.*

The substantial difference in the activation energy of the processes of surface migration and bulk flow is responsible for the difference in the temperature dependence of these processes. The ratio of the rates of both processes varies substantially with temperature. This is particularly evident in the fact that the extent to which the surface relief of pores is smoothed out differs at the same densification resulting from sintering at different temperatures and times. At low temperatures the small rate of bulk flow ensures only an insubstantial reduction in the volume of pores, but the smoothing out of the surface is substantial because of the relatively high rate of surface migration of the substance. Because of the higher activation energy, with an increase of temperature the rate of bulk flow increases far more than the rate of surface migration, and reduction in the volume of pores occurs with less smoothing out of the surface of pores. It follows that at low temperatures the smoothing out and the reduction of the surface may be considerable, with very little change in the volume of pores.

The change in surface relief and simplification of the shape of interconnected pores have a substantial effect on the gas permeability of sintered bodies. From the change in the volume of pores and the gas permeability (in the presence of interconnected pores) it is possible to judge the relative development of the processes of surface migration and bulk flow. In the tests we made

---

*In sintering of crystalline bodies with a high vapor tension at a temperature close to the melting point it is possible for the substance to be transferred by means of the gaseous phase (vaporization-condensation). This process, like surface migration, can lead only to a change in the shape of pores, with no change in the volume of pores.

the gas permeability was measured by means of a simple device making it possible to determine the amount of gas passing through a porous body in a given length of time at a constant difference in pressure. The difference in pressure amounted to 8 or 16 mm Hg, and the measurements were made by means of an oil manometer. The samples were disks 18 mm in diameter and 7 mm thick, which were mounted in a conical glass tube by means of a rubber sealing ring tightly adhering to the cylindrical surface of the sample. The difference in pressure was kept constant on both sides of the disk. The amount of gas passing through the body (air) was measured by means of a flowmeter.

The samples investigated had interconnected pores. On submersion of the sintered bodies in liquid paraffin some 97-99% of the volume of pores was filled with paraffin.

Porous bodies obtained by pressing of copper and nickel powders were investigated.

The results of the investigation of the gas permeability of sintered bodies are given in Table 18, where the change in the gas permeability is characterized by the ratio of gas permeabilities after and before sintering: G_s/G_p. At high sintering temperatures (950°C) there is a sharp reduction of the gas permeability along with reduction in the volume of interconnected pores, which would result from narrowing of the channels due to the substantial development of bulk flow. At moderate temperatures the gas permeability remains almost unchanged in some cases, despite the substantial reduction in volume of pores (sintering of nickel at 850°C, for example – see Table 18). The effect of contraction of the channels is compensated under these conditions by smoothing out of the surface. At low temperatures the gas permeability increases substantially despite some reduction in the volume of pores. In this case the gas permeability is affected mainly by surface migration of the material, leading to smoothing out of the surface and simplification of the shape of interconnected pores.

At moderate temperatures there is often some increase of gas permeability at the beginning of sintering, which then changes to a steady decline. With a complex surface relief of pores at the beginning of sintering the large gradients of capillary pressures lead to rapid smoothing out of the surface, with bulk flow predominant. This effect is observed only in the beginning of sinter-

TABLE 18. Change in the Volume of Pores and Gas Permeability After Sintering of Copper and Nickel under Various Conditions

Powder	Sintering time, min	Change in volume of pores v_s/v_p after sintering at (°C):						Change in gas permeability G_s/G_p after sintering at (°C):					
		450	550	650	750	850	950	450	550	650	750	850	950
Copper*	5	—	—	—	—	0.632	0.473	—	—	—	—	1.12	0,47
	30	—	—	0.965	0.659	0.511	0.383	—	—	1.16	1.19	0.94	0.40
	180	0.935	0.825	0.709	0.557	0.378	0.250	1.13	1.31	1.26	0.94	0.59	0.14
	540	0.924	0.780	0.673	0.466	0.311	0.212	1.17	1.32	1.34	0.77	0.30	0.05
	1800	—	0.759	0.651	0.354	—	—	—	1.40	1.37	0,58	—	—
Nickel†	5	—	—	—	0.638	0.537	0.359	—	—	—	1.12	1.03	0.55
	30	—	0.925	0.797	0.625	0.516	0.309	—	1.13	1,27	1.12	1.02	0.51
	180	0.996	0.911	0.749	0.617	0.482	0.259	1.06	1.22	1.25	1.11	0.98	0.30
	540	0.993	0.899	0.734	0.601	0.452	0.210	1.09	1.39	1.28	1.15	0.94	0.27
	1800	—	0 879	0.722	0.588	—	—	—	1.34	1.25	1.06	—	—

*Apparent density 2.62 g/cm³, compacting pressure 600 kg/cm², d_p = 4.45-4.50 g/cm³, G_p = 1.80-1.90 cm³/min.

†Apparent density 0.94 g/cm³, compacting pressure 780 kg/cm², d_p = 3.13-3.22 g/cm³, G_p = 1.25-1.36 cm³/min.

ing, since the sharp difference in capillary pressures is rapidly eliminated by the development of bulk flow; later, bulk flow results mainly in a reduction of gas permeability due to the contraction of channels of interconnected pores.

The differences in the surface and volume of pores in sintering under different conditions becomes more obvious when one compares the gas permeability of bodies in which equal reductions in the volume of pores are obtained at different temperatures. From a large number of sintered bodies, samples were selected with similar values of density but obtained under different conditions – after long sintering (30–33 h) at low temperature (650°C) and short sintering at a higher temperature (5 min at 850°C). Short sintering at 850°C for nickel as well as for copper resulted in a slight increase of gas permeability (by virtue of the effect mentioned earlier). Long sintering at 650°C caused a more substantial increase of the gas permeability in both cases (Table 19). The results of the experiment indicate that at low temperatures with equal reductions in the volume of pores there is more complete smoothing out of the surface of pores. The experiments also comfirm the possibility of a substantial reduction in the surface of pores with a small change in the volume of pores and indicate the lack of any direct relationship between the reduction in the volume and the surface of pores.

The experiments, however, were made in 1952 [55] and do not show fully the change in the surface and volume of pores, since the conclusions are based on the gas permeability, which is a complex function of the surface relief and volume of pores. There-

TABLE 19. Change in Gas Permeability at Same Change in Density after Sintering under Different Conditions

Powder	Sintering temperature, °C	Sintering time, min	d_p, g/cm³	d_s, g/cm³	Increase in gas permeability after sintering, %
Copper	650	1800	4.68	5.64	47
	850	5	4.68	5.68	13
Nickel	650	1980	4.47	5.41	38
	850	5	4.44	5.43	12

fore, later experiments were made with more accurate methods of determining changes in the surface of interconnected pores (by measuring the adsorption of methanol).

The amount of methanol adsorbed was determined at a methanol vapor pressure of 45-50 mm Hg. The division of the isotherm of methanol adsorption by metal powders in the range of these pressures is close to the inflection point on the isotherm corresponding to a monomolecular coating of the surface of the adsorbent. Therefore the amount of methanol adsorbed at this pressure can be taken as approximately proportional to the surface. Since the volume of the test tube in which the adsorbent was placed was limited, and with a low value of the specific surface it was impossible to increase the amount of adsorbent without substantial alteration of the apparatus, the accuracy of the measurements depended on the specific surface. With adsorption of order 1 mg/g the accuracy in determining adsorption was ±8% with 95% confidence and about ±15% at 0.1 mg/g. In the measurements given below the adsorption is characterized by the amount of methanol adsorbed, in mg/100 g adsorbent (mg/g × 100).

The volume of pores and the amount of methanol adsorbed, characterizing the relative value of the surface of pores, were determined after short and long sintering of copper and nickel compacts at different temperatures. The samples compared had the same porosity after sintering (brief high-temperature or prolonged low-temperature sintering). In order to reduce the development of processes occurring at low temperature as much as possible, high-temperature sintering was conducted with heating to sintering temperature at the highest possible rate (in 1-2 min). Heating was stopped when the given temperature was reached (otherwise densification was too high and it was impossible to obtain the same densification at low temperature). Twenty cylindrical samples were sintered under previously selected conditions ensuring equal densification. In determining the adsorption all twenty samples were placed in the adsorption apparatus, which ensured satisfactory accuracy of the measurements. The relative reduction of the surface, S_s/S_p, was found from the adsorption before and after sintering on the basis of the assumption that the adsorption is proportional to the surface of interconnected pores. The gas permeability and density after sintering were also measured, and the relative change in the volume of pores v_s/v_p and the relative change in the gas permeability G_s/G_p were determined from these data.

The results after short and long sintering of copper and nickel compacts are given in Table 20. In both cases the surface of pores is reduced far more after prolonged low-temperature sintering than with an equal change in the volume of pores after high-temperature sintering. After short sintering of nickel compacts at 680°C the volume of pores decreased more than the surface of pores (the volume and surface decreased 35% and 22%, respectively). The narrowing of channels with some leveling of the surface relief led to a reduction of gas permeability. After prolonged sintering at 600°C the volume of pores decreased less than the surface (the volume by 34% and the surface by 43%). Considerable leveling of the surface relief not only compensated the narrowing of the channels but also led to an increase of the gas permeability.

TABLE 20. Gas Permeability G_s/G_p and Surface of Pores S_s/S_p in Copper and Nickel Sintered under Different Conditions with Approximately the Same Change in Volume of Pores

Powder	Treatment	v_s/v_p	Methanol adsorbed, mg/100 g	S_s/S_p	Gas permeability, cm³/min	G_s/G_p
Copper	Compacted*	1.00	9.93	1.00	1.39	1.00
	Held 48 h 550°C	0.810	0.85	0.086	1.66	1.19
	Heated to 800°C in 110 sec	0.826	1.51	0.152	1.49	1.07
Nickel	Compacted†	1.00	27.1	1.00	0.63	1.00
	Held 34 h 600°C	0.665	15.4	0.57	0.91	1.43
	Heated to 680°C in 67 sec	0.654	21.0	0.78	0.45	0.71

*Relative density 0.51.
†Relative density 0.41.

It should be noted that in experiments with nickel compacts the difference between the temperatures of high- and low-temperature sintering amounted to only 80°C. Such a small difference in temperature necessitated a difference in sintering time of approximately 2000 times in order to attain equal densification. The results of the experiments also show the extremely rapid deceleration (decrease in rate) of densification at low temperature, which is characteristic of nickel powders obtained by reduction of oxides in hydrogen (see Chapter IV). The change in the rates of surface migration and bulk flow with changes in temperature is quite evident in these experiments.

Measurements of the gas permeability and calculation of the surface of interconnected pores from the adsorption of methanol made it possible to determine the development of surface migration under various sintering conditions, which is one of the elementary processes of sintering that substantially changes the structure of pores but does not cause any direct change in the volume of pores. Since the surface energy is the main driving force of densification, it can be concluded that by reducing the surface of pores surface migration must have an indirect effect on densification. In particular, it can be assumed that reduction of the surface and not the change in the concentration of lattice imperfections induces changes in the porous body sintered in the unfavorable range of low temperatures that lead to reduction of densification in later high-temperature sintering. From this viewpoint, the presence of unfavorable temperatures, which was discussed in the previous chapter, could be due to the preferential development of surface migration at relatively low temperatures. This assumption may appear justified above all to investigators who consider that "the variation of shrinkage with the surface of metal powders has definitely been established" [57]. However, more detailed analysis of the effect of preliminary calcination of powders on densification during later sintering indicates that the densification rate depends to a greater extent on the ratio of the sintering temperature to the preliminary heating temperature of the compact or the powder than on the specific surface. This was confirmed by special experiments with double sintering in which the surface of interconnected pores was determined by the adsorption of methanol after presintering and the densification rate during final sintering was determined with a dilatometric apparatus. We must examine these experiments in some detail.

Since the accuracy of determining adsorption depends on the amount of material used, special cylindrical samples with a large volume were prepared, which were sintered simultaneously under various conditions with rectangular samples intended for dilatometric studies. In selecting conditions for measuring the densification rate during final sintering it was kept in mind that the densification rate decreases rapidly with time and that the rate of decline differs at different temperatures. Therefore, to avoid the effect of additional factors (including changes in the densification rate in relation to the conditions of heating to the given temperature) the dilatometric determination of the rate of shrinkage was conducted with extremely rapid heating (90-120 sec from the beginning of heating). The average shrinkage rate in the process of heating in a given temperature was determined: 544-640, 625-730, and 750-865°C. To attain the desired experimental conditions the dilatometric tube was placed in the furnace previously heated to temperatures of 800, 900, and 1000°C. The heating rate in measurements of shrinkage in different temperature ranges differed somewhat and amounted to 190, 210, and 230 deg/min, respectively.* The rate of reduction in volume of pores was determined from measurements of shrinkage. Active nickel powders obtained by reduction of oxides in hydrogen at 450°C were used in these experiments. The results, given in Table 21, show that an increase of the presintering temperature, inducing some change in the surface of pores, leads to a far from identical change in the rate of reduction in volume of pores in different temperature ranges. An increase of the presintering temperature from 400 to 500°C, inducing a reduction of adsorption and, consequently, the surface by 43%, decreased the rate of reduction in volume of pores by a factor of 4.2 at a final sintering temperature of 544-640°C, by a factor of 1.9 at 625-730°C, and by only 10% at 750-865°C. On further increase of the presintering temperature, with reduction of the sur-

*It is not difficult to measure the shrinkage rate at different rates of increase in temperature. But the time for heating to a given temperature range would differ, which can only augment the distorting influence of nonidentical heating conditions up to the beginning of measuring the shrinkage.

TABLE 21. Effect of Presintering Temperature on Relative Surface of Pores and the Rate of Reduction in Volume of Pores During Subsequent Sintering of Nickel*

Presintering temperature, °C	Reduction in volume of pores after presintering, v_s/v_p	% interconnected pores	Methanol adsorbed, mg/100 g	Reduction of surface of pores after presintering,† S_s/S_p	Rate of reduction in volume of pores $dv/d\tau \cdot v$, min^{-1}, during final sintering at different temperatures (°C)‡		
					545—540	625—730	750—865
—	1.0	95.9	55	1.00	0.187	0.231	0.298
300	1.0	95.5	55	1.00	0.192	0.267	0.312
400	0.968	96.7	48	0.87	0.147	0.255	0.330
500	0.726	98.2	27	0.49	0.035	0.138	0.296
600	0.640	98.7	21	0.38	0	0.019	0.223
700	0.515	99.3	13	0.24	0	0	0.105
800	0.328	99.1	9	0.16	0	0	0

*Nickel powders obtained by reduction of oxides in hydrogen at 450°C, apparent density 0.86 g/cm³, d_p = 3.45-3.55 g/cm³.

†Surface of pores in compact taken as unity.

‡Rate of reduction in volume of pores measured during continuous increase of temperature in time intervals of 90-120 sec from beginning of heating.

face of pores by a further 21%, the rate of reduction in volume of pores at 544-640°C decreased almost to zero (it was so low that it could not be measured); at 625-730°C it decreased by a factor of 7, and at 750-865°C by only 25%.

Thus, these investigations as well as other experiments (when they did not amount to a simple comparison of the surface of pores and shrinkage, which under certain sintering conditions are interdependent) have shown that there is no definite relationship between the densification rate and the specific surface of the sintered body or the original powder. At the same time, similar experiments show another relationship – a connection between the densification rate in high-temperature sintering and the ratio of the presintering temperature (or preliminary calcination of the powder) and the final sintering temperature.

The variation of the rate of reduction in volume of pores with the sintering conditions, as in the variation of shrinkage with time, is determined not by the change in the surface of pores but by other laws. These laws are adequately explained on the basis of the variation of bulk flow with the concentration of imperfections in the crystalline substance being sintered. The closer the preliminary calcination temperature to the sintering temperature, the more consistently healed the defective sections during presintering and the lower the concentration of imperfections and, along with it, the densification rate at the beginning of final sintering. With an identical temperature in presintering and final sintering the densification rates at the end of presintering are approximately equal but small due to the reduction in the concentration of imperfections during presintering. However, if final sintering is conducted at temperatures below the presintering temperature (or preliminary calcination of the powder) then the densification, measured in a short time interval, is almost nonexistent regardless of the relationship with the surface of pores (Table 21).

Thus, the densification rate depends not on the specific surface of pores in compacts subjected to high-temperature sintering but on the ratio of the final sintering temperature and the presintering temperature, which can be due only to the dependence of the densification rate on the concentration of imperfections.

The results of determining the percentage of pores filled by immersion of the sintered body in liquid paraffin (i.e., interconnected pores) are also given in Table 21. Since the original powder consisted of particles with a complex shape, around 4 vol.% of the pores in the compact were not filled with paraffin. With increasing sintering temperatures the percentage of pores filled with paraffin increased regularly, although the volume of pores decreased substantially at the same time. The increase in the percentage of pores filled with paraffin was due to the simplification of the shape of interconnected pores and the disappearance of small protrusions and depressions. On the whole, these observations are in full agreement with our conclusions based on measurements of gas permeability.

It was found experimentally that the densification rate increases in all temperature ranges after calcination of the samples at 300°C. Heating of compacts at temperatures below the temperature at which the nickel powder was obtained (450°C) led to no change in the surface or volume of pores. Nevertheless, calcination had no noticeable effect on the densification rate during subsequent high-temperature sintering. One of the possible explanations for the higher densification rate after low-temperature calcination is additional reduction of oxides contained in the nickel powder (the oxygen content of the nickel powder was around 0.4%). With very rapid heating in dilatometric measurements the oxides in the compact are retained despite the hydrogen atmosphere and, separating the metal powders, they prevent densification. Preliminary heating of the compact in hydrogen reduces the concentration of oxides and eliminates their inhibiting influence on densification.

Another explanation is also possible. With rapid heating of the compact to relatively high temperature the size increases somewhat due to the thermoplastic aftereffect (see Chapter II). It is possible that with rapid heating the thermoplastic aftereffect coincides with densification, reducing the latter. Preliminary heating at 300°C eliminates a substantial percentage of the strain hardening resulting from compacting and reduces the effect of thermoplastic expansion on densification during heating to a higher temperature.

The effect of these factors (impurities of reduced oxides and thermoplastic aftereffect) is possible only in the heating period at the beginning of sintering. There are no indications of such an effect in the period of isothermal sintering.

A review of the experiments making it possible to determine the relative development of bulk flow and surface migration under various sintering conditions must conclude with an examination of the results from experiments in which more precise observations were made concerning the relative development of bulk and surface transfer of the substance during sintering of compacts with substantial lattice imperfections. It has been repeatedly observed that in sintering of some powders the variation of the gas permeability differs from that described above, undergoing a drop at the beginning of sintering that later changes to a gradual increase. This phenomenon is observed in the case of powders undergoing a particularly abrupt reduction of the densification rate with time during sintering, i.e., in the case where the value of m in Eq. (III-5) has a large value.

A similar anomalous change in the gas permeability is observed in sintering of nickel powder obtained by reduction of oxides at 450°C and compacted under a pressure of 1500 kg/cm^2. The results of measuring the gas permeability, volume of pores, and adsorption of methanol after different sintering times are given in Table 22. The gas permeability first decreased, but then began

TABLE 22. Change in Volume of Pores v_s/v_p, Surface of Pores S_s/S_p, and Gas Permeability G_s/G_p with Sintering of Active Powders of Nickel*

Sintering conditions		v_s/v_p	S_s/S_p	G_s/G_p†	% interconnected pores
Temperature, °C	Time, h				
800	—	0.505	0.32	0.69	99
	0.25	0.475	0.25	0.67	99
	0.5	0.469	0.23	0.76	99
	1	0.461	0.22	0.78	100
	2	0.460	0.22	0.82	100
	3	0.458	0.21	0.83	100
	6	0.442	0.20	0.85	100
700	—	0.778	0.71	0.86	98
	0.25	0.649	0.45	0.83	99
	0.5	0.637	0.39	0.83	99
	1	0.618	—	0.95	100
	2	0.602	0.34	0.99	100
	3	0.604	0.32	0.99	100
	6	0.602	0.29	1.00	100

*Powder obtained by reduction of oxides in hydrogen at 450°C, apparent density 0.86 g/cm^3, d_p = 3.48-351 g/cm^3, G_p = 0.97 cm^3/min.
†The relative gas permeability of the compact was taken as unity.

TABLE 23. Change in Volume of Pores v_s/v_p, Surface of Pores S_s/S_p, and Gas Permeability G_s/G_p after Sintering of Nickel Powder First Calcined, with a Small Difference between the Calcination and Sintering Temperatures*

Sintering time, h	v_s/v_p	S_s/S_p	G_s/G_p	Sintering time, h	v_s/v_p	S_s/S_p	G_s/G_p
3	0.94	0.86	0.94	49	0.89	0.68	0.98
6	0.92	0.76	0.92	84	0.88	0.67	1.04
12	0.92	0.77	0.91	120	0.88	0.63	1.08
18	0.91	0.73	0.93	144	0.88	0.61	1.09
24	0.91	0.71	0.95	170	0.88	0.59	1.10
30	0.90	0.71	0.96				

*Original material was nickel powder calcined at 650°C, sintered at 700°C, d_p = 3.63 g/cm³, G_p = 23.6 cm³/min.

to increase after sintering for 15-30 min and continued to increase until the end of sintering (6 h). This variation of the gas permeability with time is probably directly connected with the change in the flow rate of the crystalline substance with time.

The flow rate, and the rate of reduction in volume of pores along with it, decreases rapidly with time because of the simultaneous development of the elimination of imperfections. Therefore the ratio of the rates of bulk flow and surface diffusion must depend not only on temperature but also the concentration of imperfections. With a large concentration of imperfections and rapid increase of temperature a very high rate of bulk flow in the beginning of sintering leads to rapid narrowing of the channels, the effect of which on the gas permeability cannot be compensated by surface migration of the substance, which cannot change the surface relief of the pores in a short time. However, after the abrupt slowdown in reduction of the volume of pores the continuing migration smooths out the irregularities in the surface and simplifies the shape of the channels of interconnected pores, which leads to a gradual increase of the gas permeability.

In the case where densification is characterized by a substantial "self-stopping"effect [i.e., coefficient m in Eq. (III-5) has a large value] but the sintering temperature only slightly exceeds the calcination temperature of the powder the reduction of the gas permeability is also replaced with an increase, but with a longer sintering time. The phenomenon is wholly retained but stretched over a longer time. The results from investigations of sintering compacts of nickel powder calcined at 650°C are given in Table 23. The sintering temperature was 50° above the calcination temperature. The densification was negligible, and after a long sintering time the reduction in volume of pores became vanishingly small. The gas permeability decreases in the first 12 h and then slowly increases. The gas permeability increases up to the end of sintering for 170 h (with interruptions for the measurements). The variation of the ratio of the rates of surface migration and bulk flow with time during isothermal sintering was also determined in this experiment. The increase of the gas permeability at the beginning of sintering was due to reduction in the volume of pores, which then slowed down due to reduction of the concentration of imperfections. With further sintering the surface migration continued, causing reduction and smoothing out of the surface of pores, which were manifest in a gradual increase of the gas permeability.

It is interesting that the increase of the gas permeability and reduction of the surface (judging from the adsorption of methanol) continued after apparent cessation of any change in the volume of pores. Evidently the rate of surface migration decreases only due to smoothing out of the surface relief and reduction of the difference in the curvature of adjacent sections of the surface. The substructural characteristic of crystalline particles, i.e., the concentration of imperfections, has almost no effect on this process (although it is not very probable that it has no effect, since imperfection emerging at the surface can change the energy condition of surface atoms).*

Thus, analysis of the experimental data leads to the following conclusions: The ratio of the rates of bulk flow and surface migration depends not only on the values of the activation energy and temperature but also on the concentration of imperfections in the crystal lattice of the material. In crystals with an imperfect lattice the ratio of the rates of these processes may differ substantially from the "normal" ratio for crystals under equilibrium conditions.

---

*Here, the author states an opinion opposite that expressed in his first publications on theoretical problems of sintering [1]. Later studies made it necessary to abandon the idea that the "energetic nonuniformity of the surface" of pores is the decisive factor.

Surface migration of the material can have a substantial effect on gas permeability only under those sintering conditions where the rate of surface migration is relatively high and the rate of bulk flow is low. When the temperature is not too high this condition is almost always realized after the concentration of imperfections is substantially decreased in the process of sintering and the densification rate becomes fairly low. If the densification rate is already low in the beginning of sintering then the gas permeability may begin to increase directly at the beginning of isothermal sintering. With a high sintering temperature the variation of the gas permeability depends mainly on the bulk flow of the material, and the gas permeability usually begins to decrease soon after the beginning of sintering.

Because of the substantial yield of the material even at a low sintering temperature when the concentration of imperfections is high, the changes in the surface and shape of pores may depend on bulk flow, with no noticeable effect of surface migration on this process (additional experimental data supporting this conclusion are given in the following chapter).

Experiments revealed no clear relationship between the densification rate and the relative value of the surface of pores. This relationship (which, in general, should exist along with the others, since the driving force of densification is surface energy) is suppressed in isothermal sintering by the stronger influence of the concentration of imperfections on the flow of material. However, the change in the geometry of pores in the process of isothermal sintering should also have some effect on the course of densification. Although the phenomenological manifestations of this relationship are not very clear in isothermal sintering of metal powders, they must be revealed in order to describe the densification process by means of kinetic equations of the elementary processes. The character of this relationshp is examined in Chapter IX.

Chapter VII

The Flow of Metal under the Influence of Surface Tension at Room Temperature

Of considerable interest among the various manifestations of the dependence of the flow rate of crystalline materials on the concentration of imperfections is the high flow rate of imperfect crystals of metals obtained by electrolysis of salt solutions at a high current density. The dendrite-like crystals obtained under these conditions undergo a change in shape under the influence of surface tension at room temperature. Observations of the changes in the shape of dendrite-like crystals and the gaps (fissures) between them can yield important additional information on the nature of elementary processes developing during sintering.

The flow of metal under the influence of surface tension at room temperature was discovered in 1952 by Garber and D'yachenko at the Khar'kov Polytechnical Institute [58]. The significance of their observations was not properly appreciated by metal physicists either here or abroad. These observations are clear illustrations of the effect of lattice imperfections on the flow rate of metals.

The powders investigated were obtained by electrolysis of salt solutions. The rapid growth of crystals at room temperature was ensured by high concentrations of imperfections. Separate sections of the contour of particles were examined by means of an electron microscope. Spontaneous changes in the shape of protrusions, closing of gaps, and disappearance of interconnected pores (visible with transillumination) were observed. A certain vagueness in the conclusions (the hypothesis of even growth of a

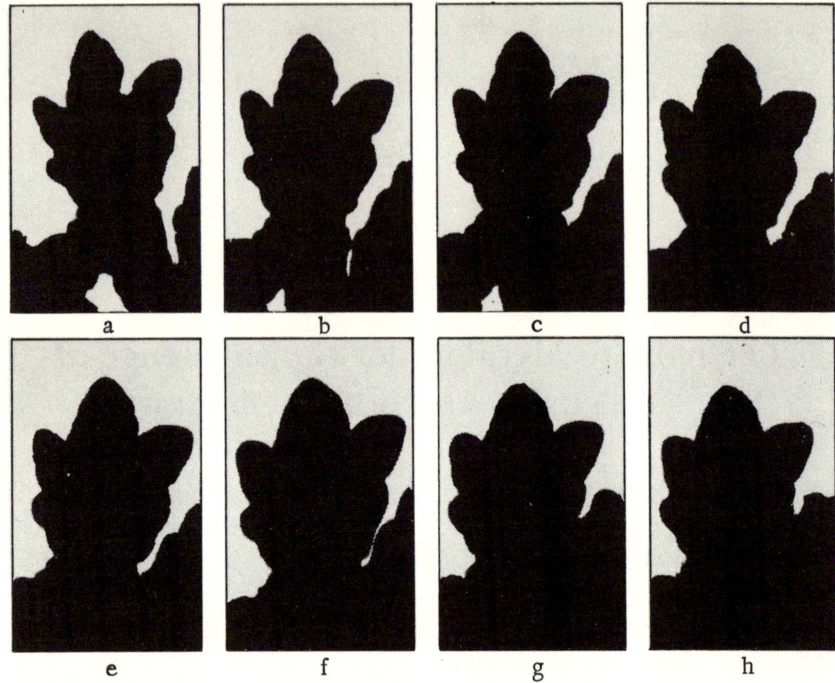

Fig. 18. Change in the shape of copper dendrite at room temperature (electron micrograph, 5000×). a) After 3 days; b) 7 days; c) 20 days; d) 28 days; e) 52 days; f) 82 days; g) 123 days; h) 162 days.

layer of substance on all sections of protrusions or pores, with uneven development of this process in time) caused the present author to repeat these experiments to determine more exactly the characteristics of flow under these conditions and to reveal any regularity in the course of the process with time.

The technique of the experiment described by Garber and D'yachenko [58] was completely reproduced in our work. Copper was obtained by electrolysis of copper sulfate at a high current density (~5 Å/cm^2). The powders were washed in alcohol and ether and placed on the specimen holder of the electron microscope. Selected sections of the contour were photographed at magnifications of 5000-10,000×. The specimens were first photographed every day, and then, after the process slowed down, at longer time intervals. Changes in the shape of the dendrite-like protrusions were observed on the surfaces of particles and fissures with a complicated profile (with protrusions on the inner surface). The specimens were examined for changes in the shape of protrusions over a period of 180 days. Several photographs characterizing the changes in the shapes of protrusions are shown in Fig. 18. Regretfully, no observations were made in the first three days because of methodological complications. However, even after three days from the time the powder was obtained the rate of change in the shape of the protrusions was still high. In time it decreased. The change in the shape of the protrusions reminded one of the fusion of an amorphous body: The length of the protrusion decreased, the width increased, the curvature of all sections of the surface decreased. These changes can be taken as obvious confirmation of the role of surface tension, which is the driving force of flow during sintering of any material, including crystalline materials.

The photographs of the protrusions were used to determine the variation in the flow rate of the material with time. In view of the specific shape of the protrusion selected, with an irregular configuration (bulges and depressions on the sides), the degree of deformation ("broadening") of the protrusion could be characterized by the change in the ratio of its cross section b measured at the base of the lateral protrusions to the height h, taken as the length of the perpendicular dropped from the tip of the protrusion to the base line. A diagram of the method by which the protrusions were measured is shown in Fig. 19 and the change in the ratio of the width to the height with time in Fig. 20. The rate of change in the shape of the protrusion decreases rapidly with time. This reduction occurs far more rapidly than that of the difference in the capillary pressures due to the change in shape. Approximate calculation of the capillary pressures at the tip of the protrusion and at the base showed that after 17 days the difference in capil-

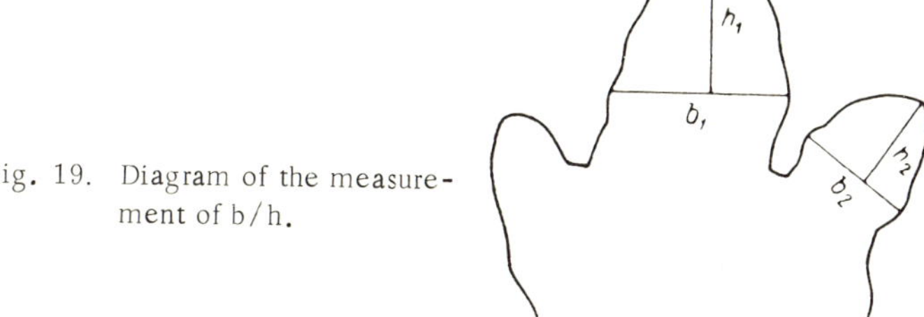

Fig. 19. Diagram of the measurement of b/h.

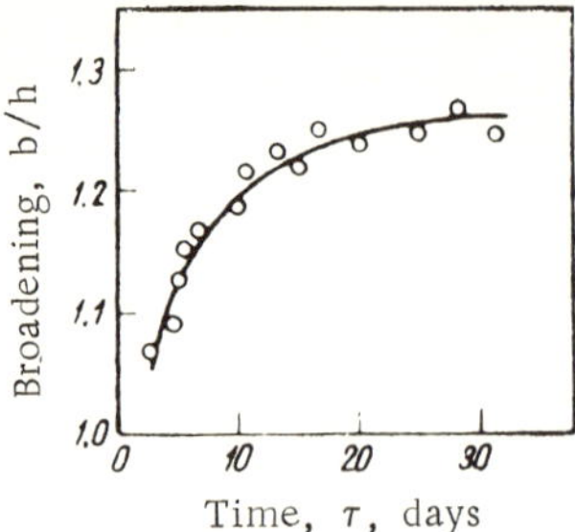

Fig. 20. Change of b/h with time.

lary pressures decreased by 20%, while the deformation rate, calculated from the change in b/h, decreased by a factor of 25.

As noted above, in these experiments the observation of the change in shape was begun three days from the time the powder was obtained. It can be assumed that in this time the flow rate had already decreased substantially. For this reason, attempts were made to observe changes in the shape of the particles near the time at which the powder was obtained. A thin copper wire (0.2 mm in diameter) was used as the electrode in electrolysis of a solution of copper sulfate. Together with the crystals formed on its surface, the wire was mounted on the specimen holder adapted for this purpose and rapidly transferred to the microscope. The electrolysis lasted 3 sec at a current density of ~5 Å/cm^2. Transfer to the electron microscope and preparation of the microscope (creating the necessary vacuum) took several minutes. The first photograph was made in 8-9 min after extracting the wire from the electrolytic bath. Subsequent photographs were made every 15-30 min.

The experiments showed a very high deformation rate of the dendrites. A change in shape can be seen in photographs taken 15 min apart (Fig. 21). Because of the complex shape of the dendrite-like formations, it was impossible to measure the deformation rate under conditions comparable with those in the preceding experiment. However, it can be assumed that the deformation rate in the later experiment was no more than one order higher than the rate first measured in the preceding experiment. The simplification of the shape of the dendrite-like protrusion is very striking in the first photograph. There is sufficient reason to assume that electrolysis produces the usual sharp-angled dendrites and that the notable smoothing out of the sharp edges and protrusions occurs as early as the time in which the sample is prepared for photographing.

After several days the group of particles in the microscope had changed so much in outline that it was impossible to recognize them. This probably explains the unsuccessful beginning of the preceding experiment – the outline of the particles changed greatly and therefore the point at which the particles were photographed was impossible to find. In subsequent experiments the specimen holder was not removed from the electron microscope.

The specimen is subjected to irradiation with electrons in the microscope, and therefore it is necessary to consider the possible effect of irradiation on the flow rate. Garber and D'yachenko [58, 59] made a special effort to observe this effect. The samples were exposed (i.e., irradiated) under ordinary conditions 10 times without interruption and then 10 times with a period of 5 min between photographs. In the second series of experiments the total intensity of irradiation was the same as in the first, but the experiments were extended in time. No changes were noted in the sizes of gaps between particles of powder in the first case, and considerable deformation was noted in the second case, when a great deal of time elapsed between the first and subsequent exposures. The experiment showed that deformation develops evenly in time and that irradiation in an electron beam has no noticeable effect on the deformation process. The conclusions drawn by D'yachenko [59] are confirmed by investigations of the influence of electron irradiation on the creep of copper [60], which revealed no noticeable change in the creep rate. In other studies [61-63] no effect of electron irradiation was noted on the creep rate of

Fig. 21. Change in the shape of dendrite-like crystals of copper in the first hour after the powder was obtained by electrolysis (7500×). a) 9 min from the end of electrolysis; b) 25 min; c) 56 min.

aluminum, nickel, or iron. In our work it was also possible to observe that the deformation rate changes evenly with time regardless of the number of times the sample had previously been exposed to irradiation (number of exposures). Evidently, the energy of electrons at the difference of potentials used in the electron microscope is insufficient for the transfer of atoms even in a nonequilibrium lattice, and the slight heating of the object that is possible has no noticeable effect on flow because of the short exposure time (3 sec).

Thus, there is sufficient reason to consider that the high deformation rate of dendrite-like crystals observed in these experiments is induced by the very high concentration of imperfections, and the change in the flow rate with time is due to the spontaneous elimination of imperfections.

There are direct experimental indications that a particle of metal with a small concentration of imperfections changes its shape negligibly even after a long time. Figure 22 shows micrographs of a protrusion on the surface of a copper particle obtained from a piece of copper by means of a needle file. Separation of a particle from the piece results in cracks in the edge, on which there are protrusions approximately the size of protrusions on dendrite-like crystals with imperfections. Photographs were made after 8 and 162 days rest at room temperature. In comparing the photographs in Figs. 18 and 22 it should be kept in mind that the photographs in Fig. 22 were made at almost double the magnification (the respective scales are shown by the length of the micron drawn on the figures).

After 162 days the protrusion shown in Fig. 22 retained its shape on the whole, although in some sections where a bulge or depression has a very small radius of curvature there are changes. It is doubtful whether flow of the metal in these sections is due to lattice imperfections. Imperfections induced by plastic deformation during separation of the filing cannot be the reason for the high flow rate. This follows, for example, from the fact that the preceding deformation does not increase but reduces the creep of copper at room temperature or slightly elevated temperature [64]. With a very small radius of curvature the capillary pressure may exceed the yield strength of copper. With a radius of curvature of $0.1\,\mu$ the capillary pressure is of order 10^8 dyn/cm^2. The yield

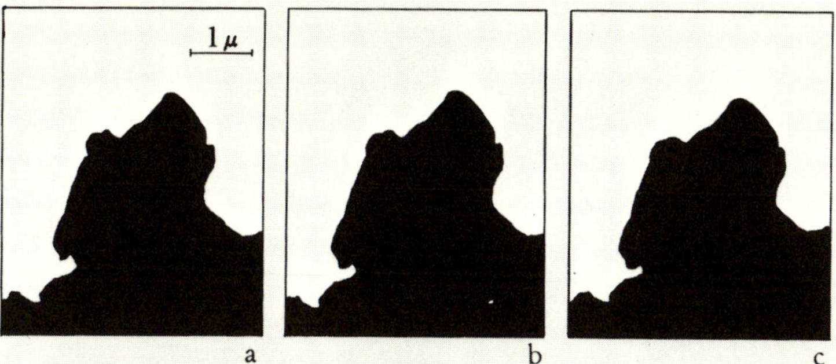

Fig. 22. Change in shape of a section from a particle of copper mechanically separated from a piece of solid metal (8000×). a) Original specimen; b) after 8 days; c) 162 days.

strength of copper is from 10^7 dyn/cm^2 (perfect single crystal) to 10^9 dyn/cm^2 (imperfect single crystal and polycrystalline copper). Plastic deformation can evidently occur in sections with a large curvature with a small concentration of imperfections. However, it is not clear why, if the yield strength is exceeded, the deformation develops so slowly or whether the dislocation mechanism can be responsible for the development of such slow deformation.

Without going into details,* let us note that plastic deformation, while possible, develops only in individual sections of particles delimited by a surface with a very small radius of curvature.

Thus, investigations showed a radical difference in the change in the shape of dendrite-like defective crystals and particles of copper separated mechanically from a piece of cast metal: In the first case deformation encompasses the entire bulk of the particle, while in the second case it involves only individual sections near surfaces of large curvature. Evidently, deformation of the dendrite is due to the type of flow for which the rate depends on the concentration of impurities formed in the growth of the crystal.

The change observed in the shape of dendrites obtained by electrolysis is due to bulk flow of the material. This conclusion can be drawn from the following observations. The length of the protrusions decreases and the width increases with the contrac-

*The possibility of the dislocation mechanism acting will be considered in the following chapter.

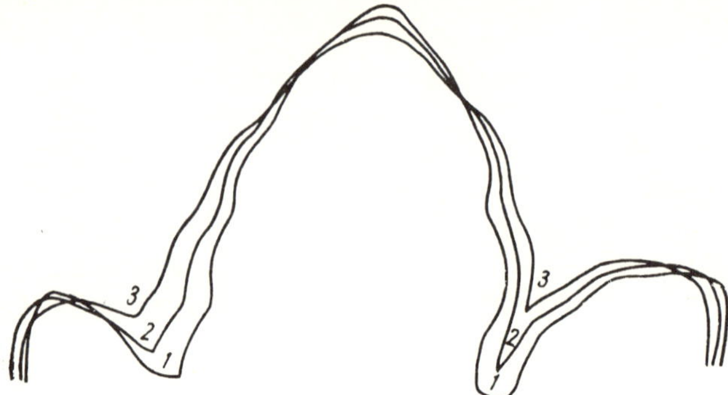

Fig. 23. Change in the profile of a protrusion from a particle of copper obtained by electrolysis after rest at room temperature. 1) After 3 days; 2) 15 days; 3) 60 days.

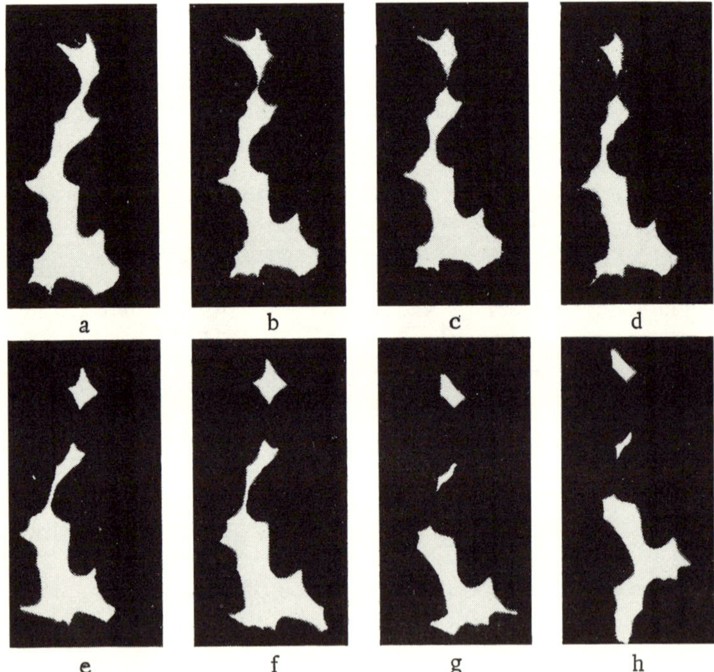

Fig. 24. Change in contour of fissure in a spongy particle of copper obtained by electrolysis (5000 ×). a) After 5 days; b) 11 days; c) 36 days; d) 57 days; e) 82 days; f) 104 days; g) 120 days; h) 162 days.

tion of small depressions and bulges on the sides of the protrusions. If the change in shape were due to surface migration of the material then this process would lead first of all to smoothing out of the fine surface relief – filling of small depressions and disappearance of small bulges – and only then affect the shape of the protrusion as a whole. The broadening of the protrusion with the contraction of bulging sections on the sides would be impossible in this case (growth of the layer in the area of the bulge due to surface migration is unjustified in terms of energy). Meanwhile, the small irregularities in the form of bulges and depressions gradually even out but do not disappear altogether even after considerable change in the shape of the protrusion (Fig. 23).

Surface migration cannot play any notable role in the change of the shape because of the very small rate of this process at room temperature. A high concentration of imperfections ensured the possibility of bulk flow under conditions in which there was almost no surface migration.

Thus, here again we encounter the "anomalous" relationship between the rates of surface migration and bulk flow induced by a high concentration of imperfections. For this reason, preferential development of bulk flow is possible at low temperature, but not surface migration, as follows from the ratio of the activation energies (similar observations were given in Chapter VI for sintering of active powders).

The pattern of the change in the shape of fissures (Fig. 24) is similar in many ways to the change in the shape of protrusions. The width of the fissure in any given section first decreases rapidly, and then the process slows down. However, the change in the size of individual sections of the fissure over a long period of time is complicated by secondary phenomena, the analysis of which gives additional data concerning the effect of the changing geometric characteristics of the surface on the reduction of pores.

The slowdown of the rate of contraction of fissures is due both to a change of flow because of reduction in the concentration of imperfections and reduction of capillary pressures due to a smaller curvature of the surface near the contacting sections (between arms of dendrites or crystals of loose particles). From the analysis of the effect of these factors on the change in the shape of protrusions it is evident that the first factor is also predominant here – the change in the concentration of imperfections (the "substructural factor"). However, with reduction of slit-like pores it is possible for encounters to occur between protrusions on opposite sides of the pore (fissure) and a new contact can be formed. This may substantially change the relative effect of the geometric and substructural factors on the rate of contraction of the pore near the new contact.

Some sections of the visible contour of a fissure usually lie in different planes. In most cases there is only an apparent connection of the opposite sides of the fissure. Protrusions, without touching, fall one behind the other, making it possible to look through the fissure. The development of a new contact can be observed infrequently. However, this phenomenon does occur, which can be judged from the change in the width of neighboring sections of the fissure at the time a new contact is formed. If one follows (Fig. 25) the change in the narrow upper and middle sections of the fissure, measuring the distance from the bulge to the opposite side, one finds that after contact of the upper section the shrinkage of the middle section is accelerated (see Fig. 25b, where the changes in the upper and middle sections of the fissure are shown in a graph).

With the formation of a new contact a new region appears with a negative capillary pressure, accelerated influx of the substance into the zone of the new contact, and joining of the adjacent sections of the fissure. However, the effect of the new contact on the change in the dimensions of other sections of the fissure is

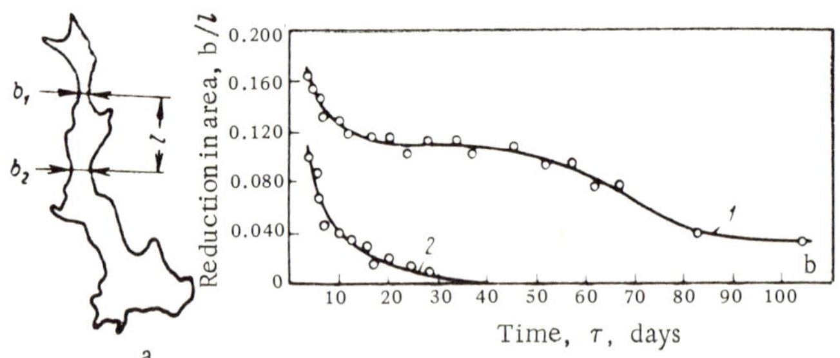

Fig. 25. Diagram of measurements of width in upper and middle sections of fissure (a) and plot of reduction in area of these sections with time (b). 1) Middle section; 2) upper section.

more complex. The new forces coming into play disrupt the regular reduction in separate sections of the pores, causing a mutual shift of crystallites and "bending" (turning) of large sections of the surface of pores. Sections of the fissure far from the new contact may even become wider for this reason. In some time interval they begin to be narrower again. This can be seen from the change in width of the lower section of the fissure in Fig. 24. Obviously, this can also occur in the case where the new contact is not visible in the opening of the fissure. It must be assumed that the irregular variation in the width of fissures and sizes of pores observed by Garber and D'yachenko [58] (i.e., the visible contours of through holes) with repeated replacement of narrowing by broadening must be explained by the periodic occurrence of new contacts, inducing accelerated joining of some sections and broadening of other sections of fissure-like pores. However, from our observations the protrusions with simple contours change shape quite regularly with a continuous decrease in rate.

On the whole, the observed changes in the shape of protrusions and fissures can serve as an illustration of the effect on the densification process (or change in shape of particles) of the basic elementary processes – bulk flow under the influence of capillary forces and elimination of imperfections, leading to reduction of the flow.

These experiments also make it possible to judge the extent of the effect of imperfections on the flow of a crystalline substance with a highly imperfect lattice. This estimate may be of considerable importance in developing theoretical concepts explaining this effect.

The acceleration of the flow of metal due to imperfections can be estimated by measuring the deformation rate of the protrusion with time (or deformation after a given time) and comparing these values with those calculated for a crystallite with the same geometrical dimensions but a perfect lattice. The deformation of a crystallite with an equilibrium lattice can be calculated from the results of creep experiments with thin wires under the influence of forces similar in magnitude to those resulting from surface tension. These experiments were conducted with wires of different metals at relatively high temperatures. Usually the flow rate was found to be proportional to the load, which made it possible to characterize the flow of metal at a given temperature by a constant coefficient of viscosity. The activation energy of flow was determined from the viscosity coefficient determined at low temperature; it was usually close to the activation energy of self-diffusion for the given metal. The latter fact was regarded as an indication of the diffusion mechanism of flow at high temperature and small load. Similar studies were made by Udin, Shaler, and Wulff [65]. They found that the variation of the viscosity coefficient η (reciprocal of flow) of copper with temperature is expressed by

$$\eta = 130 \cdot \exp \frac{59{,}000}{RT} \text{ P.} \qquad \text{(VII-1)}$$

With this equation the results from measuring the rate of flow can be extrapolated to room temperature without difficulty, i.e., calculate the viscosity coefficient and find the rate or magnitude of deformation of a cylindrical sample with a diameter equal to the cross section of the protrusion, the change in the dimensions of which is used for calculating the deformation rate with a defective lattice. By this means it is possible to determine the deformation rate of a protrusion that would occur if the deformation rate at room temperature did not differ from the creep mechanism of copper at 700-900°C (we are still speaking of creep under the influence of small forces with the same order of magnitude as surface tension).

Determining the extent of deformation from the change in the shape of the protrusion was complicated somewhat by the inconsistent magnification of the electron microscope when the micrographs were made after different time intervals. For this reason the shape of the protrusion after different holding times was determined from the ratio of its cross section to height (the method of measuring these values was described above). For cylinders of length l and diameter d, with constant volume V, relationship $l = f(d/l)$ is represented by

$$l = \left(\frac{V}{\pi}\right)^{\frac{1}{3}} \left(\frac{d}{2l}\right)^{-\frac{2}{3}}, \qquad \text{(VII-2)}$$

while the ratio of lengths l is related to the change in the ratio

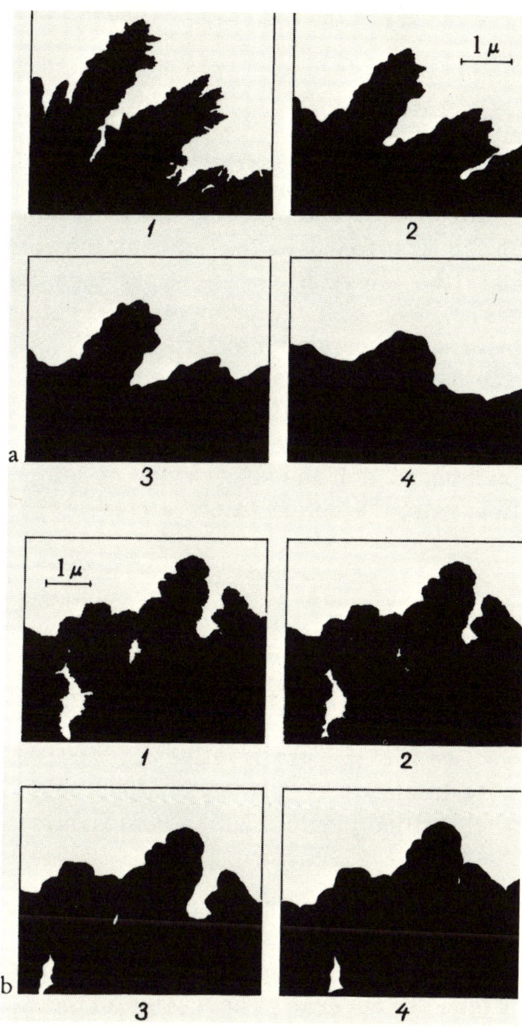

Fig. 26. Change in contour of dendrite-like crystals of silver (a) and gold (b) obtained by electrolysis, after holding at room temperature. 1) After 0.5 h from completion of electrolysis; 2) after 4 days; 3) 40 days; 4) 80 days.

d/l at constant V by the relationship

$$\frac{l_2}{l_1} = \left(\frac{d_1 l_2}{l_1 d_2}\right)^{\frac{2}{3}}. \qquad \text{(VII-3)}$$

For a regular finger-like protrusion (Fig. 18) $d_1/l_1 = 1.08$ after 3 days, while $d_2/l_2 = 1.23$ after 18 days. Since the deformation is expressed as $\varepsilon = l - l_2/l_1$, then, using (VII-3), we have $\varepsilon = 0.085$ (after 15 days). The average deformation rate for 15 days is 6×10^{-3}. In the first few days the deformation rate is much higher: 2×10^{-2} [the deformation rate was determined from the slope of $d/l = f(\tau)$, Fig. 20]. To determine the deformation rate of annealed metal with the same dimensions and shape of the protrusion as in the particle we find the value of the viscosity coefficient at room temperature by Eq. (VIII-1). At $T = 300°K, \eta = 9 \times 10^{44}$ P. With reduction of the length of the wire due to surface tension, and because of the small value of deformation and the constant flow rate, the deformation rate equals

$$\dot{\varepsilon} = \frac{\sigma}{6\eta r}, \qquad \text{(VII-4)}$$

where σ is surface tension and r is the radius of the wire.

For our case σ (surface tension of copper) is equal to 1370 dyn/cm, $r = 6.5 \times 10^{-5}$ cm ($0.65\,\mu$), and $\eta = 9 \times 10^{44}$ P. Making the calculations, we find that the relative deformation of the protrusion under the influence of surface tension would be 3×10^{-34} per day for a crystalline particle with a comparatively perfect lattice.

The deformation rate of a protrusion of a particle with an imperfect lattice proved to be $6 \times 10^{-3}/3 \times 10^{-34} = 2 \times 10^{31}$ times higher than the calculated deformation rate of the same protrusion with a perfect lattice. Several inaccuracies in the calculation, particularly the unaccounted-for effect of the diameter of the wire on the viscosity coefficient [66] and the inaccuracy associated with extrapolation of the temperature dependence far beyond the temperature range in which the dependence was established, could change the value characterizing the ratio of the rates by one to two orders, which does not change the general estimate of the effect of imperfections on the flow rate. The calculations lead to the following conclusion. If the same mechanism that controls high-temperature creep were acting at room temperature then the creep rate would be 10^{30} times smaller than the experimentally determined flow rate for a defective crystal.* Thus, a large concentration of imperfections induces acceleration of flow, expressed as a number of astronomical order. However, it is doubtful whether one can speak of the influence of imperfections on the type of flow that controls high-temperature creep. It is more likely that deformation of a defective crystal at room temperature is due to a special mechanism of flow.

Similar studies were recently made with defective crystals of silver and gold obtained by electrolysis of solutions of their salts at room temperature. As in the case of copper crystals, dendrite-like crystals of these metals rapidly change their shape at room temperature (Fig. 26). This same phenomenon was noted earlier for defective crystals of iron.† Thus, the ultrahigh flow in defective crystals is not due to the particular nature of the metal but the inherently imperfect crystals of various metals.

*Let us note that extrapolation of the deformation rate to room temperature from any of the data on high-temperature creep of copper at small loads also leads to this conclusion. In all cases, the calculation of the deformation rate at room temperature gives very small values of the rate that are of the same order, since the activation energy of creep approaches the activation energy of self-diffusion under these conditions.

†S. S. D'yachenko. Electron Microscopic Investigation of Sintering of Metal Powders [in Russian]. Dissertation, Khar'kov Polytechnical Institute (1952).

Chapter VIII

Phenomenology of Sintering and Modern Theoretical Concepts

For a quantitative estimate of the effect of the elementary processes on densification it is first necessary to describe them quantitatively as independent processes, after which one can attempt to establish (quantitatively) their interdependence and their effect on densification. The problem would be completely solved if it were possible to describe these processes on the basis of modern physical concepts of lattice imperfections and their part in the flow of a crystalline substance. However, as will be indicated below, the attempts made up to the present time have not been successful – at best they give some qualitative explanation, but no quantitative description.

Another approach to the problem of explaining the nature of the elementary processes determining the course of densification is also possible – to reveal the phenomonological characteristics of the processes and describe them by means of general considerations on the kinetics of chemical reactions, which, as is well known, are applicable to other processes developing in the crystal lattice of a solid. Such an analysis bypasses the question of the atomic mechanism of the processes and thus is not a complete solution of the problem, but it permits the essence of the processes to be explained at the level of the phenomenological theory. Furthermore, a fairly complete description of the processes that develop during sintering can later serve as the basis for a more rigorous theory employing modern physical concepts.

As already noted, phenomenological analysis of kinetic laws is justified by the absence of any solution to the problem of kinetics on a more rigorous physical basis. Let us examine this point in more detail.

Although the phenomenology of sintering gives a satisfactory basis for considering that the effect of lattice imperfections on the flow rate of the crystal determines the variation of sintering with time, a good deal of work has been done in which the kinetics of densification was considered without regard for any connection with the change in the concentration of imperfections, and the drop of the densification rate with time was attributed to the effect of other factors.

Some authors [5, 67-69] explain the drop of the densification rate with time by the reduction of capillary pressures below the yield strength due to the disappearance of sections with a small radius of curvature from the surface, which leads to a change from rapid plastic deformation to slow viscous flow.

This viewpoint was criticized by Fedorchenko and Andrievskii [54], who pointed out that these concepts are in contradiction to the high densification rate of active powders with imperfect crystals. A high concentration of imperfections can only increase the yield strength, not lower it. Furthermore, there are doubts that the capillary pressure can exceed the yield strength in volumes embracing a substantial part of the sintered body. The calculations given in [54] refute such a possibility. Herring [70] indicates that although some local values of the capillary pressure may exceed the yield strength, the contribution of plastic deformation to the overall densification process cannot be substantial, since the formation exceeding the yield strength is possible only at the tip of the contacts and in fissures with a radius of curvature of order 10^{-5} to 10^{-4} cm. Such fissures disappear rapidly at the beginning of sintering. If plastic flow generally has an effect on densification (which is most probable for fine powders with a perfect lattice) then its effect is limited by the period of the existence of thin fissures, which disappear before the beginning of isothermal sintering. The participation of plastic deformation in isothermal densification is exceedingly improbable.

Several investigators have connected the drop of the densification rate with the simultaneous grain growth. Herring's theory of diffusion flow gives some basis for it [71]. The inverse relationship between the rate of diffusion flow and the square of the average grain size postulated by this theory was used by Samsonov and Koval'chenko [37, 72] to describe the variation of densification during hot pressing.

Coble [31] used approximately the same method to deduce the kinetic law of densification. The size of pores was associated with the grain size on the basis of the geometric structure and dependence of the rate of closing of pores on the size in conformity with the diffusion theory. In this case the variation of grain size with time is described by the empirical relationship $L \sim \tau^{1/3}$. This leads to a kinetic equation similar to (III-18), the validity of which is discussed in Chapter III.

Without going into the details of the derivation, let us note that the assumption of a relationship between the flow rate and the grain size must be rejected immediately for the case of an abrupt change in the flow with no noticeable grain growth. This applies in particular to the spontaneous change in the shape of particles obtained by electrolysis at a high current density. The copper powder obtained in this manner shows dotted lines on the powder patterns and hence it follows that the size of the crystals in the powder is at least $10\,\mu$. Since a protrusion is far smaller ($\sim 1\,\mu$), and also because of its origin (a smoothed out arm of a dendrite), there is sufficient basis for the assertion that the change in the shape of protrusions described in the previous chapter characterizes the change in the time of flow of the monocrystalline section of the particle. Thus, the change in flow cannot be due to grain growth.

It should also be noted that intense grain growth does not usually occur before the development of densification, particularly in bodies of low density; one can rather conclude from the experiments that densification precedes grain growth. It is precisely at low temperatures, where densification is negligible, that one finds preferential development of those changes in the submicrostructure which substantially lower the capacity for densification in subsequent high-temperature sintering. It follows from this that there is no basis for considering grain growth to be the main factor in the decline of densification. As for the mechanism of low-temperature flow (near room temperature), the slowing down of the process is certainly not due to grain growth.*

*Further discussion of this question can be found in Chapter X.

Some decline of the densification rate with time may be due to the higher effective viscosity of the porous body because of the smaller number and size of pores. However, this effect, as shown by several investigators [4, 72], is inadequate to explain the decline of the densification rate. These investigators conclude that the theoretical and experimental data will agree only by taking into account the change in viscosity with time due to the changes in the viscosity with changes in the substructure (or structure) of the crystalline substance; this agrees with the present author's opinions, expressed earlier [55, 56].

The kinetic equations given in [31, 37, 72] were derived by using a model of a sintered body with equidimensional pores. It would be more valid to use a model of a viscous porous body with pores differing in size. In this case the change in the densification rate with time would be due to rapid disappearance of small pores, which are eliminated at a higher rate. However, as shown by calculations in Chapter V, the difference in size of pores (both closed pores and cylindrical channels) cannot be the reason for the vast difference in the densification rate at the beginning and end of sintering that is observed in experiments.

Thus, without use of assumptions concerning the effect of lattice imperfections on the flow rate it is impossible to explain the rapid decline of the densification rate during the period when interconnected pores exist. At the same time, there are substantial difficulties in calculating the effect of imperfections. Attempts to substantiate this effect still have not led to a theory capable of providing a quantitative description of the effect of imperfections on the flow rate.

One of the first such attempts was made by Pines. The diffusion mechanism he proposed for closing of pores in 1946 was later supplemented with the assertion that the diffusion flow controlling this process increases proportionally with the vacancy concentration, which far exceeds the equilibrium concentration in defective crystals [32]. The validity of these concepts, however, is disputable. The objections, presented earlier by the present author [73], are basically as follows:

1. The movement of "voids" within a porous body cannot change its total volume. In a body with interconnected pores the outer surface of the body constitutes an exceedingly small per-

centage of the surface of pores and the probability of a vacancy emerging on the outer surface is therefore exceedingly small. If we take into consideration, however, that both the outer and inner surface has identical (or almost identical) relief then it becomes obvious that there are no factors that would compel vacancies to move along the chain of particles to the outer surface of the body.

With interconnected pores the sintered body undergoes densification evenly throughout the volume. This is confirmed by the absence of any preferential flow of vacancies to the outer surface of the body. In a body with interconnected pores the outer surface and the surface of the pores serve as a vacancy sink. Vacancies formed in concave sections will tend to the nearest convex section of the surface within the limits of the given pore or the next pore. This process will induce no change in the volume of pores, but will lead, as in surface migration of atoms, to leveling of the surface and rounding off of the pores.

A refinement of this mechanism proposed by Geguzin [74, 75] – the preferential movement of vacancies along interblock or intercrystalline boundaries – gives no explanation for the densification of a porous body with interconnected pores. The rapid shrinkage of a single pore in an imperfect crystal or polycrystalline body may be due to rapid diffusion of vacancies along interblock or intercrystalline boundaries, but the application of this mechanism to a group of particles with interconnected pores gives no explanation of the densification process. The migration of vacancies along block or grain boundaries, as in the case of diffusion within a particle, can induce the departure of atoms only from one section of the surface, with an equal buildup of the lattice in another section of the surface of the pore. This process cannot induce any change in the volume of pores.

2. The application of the diffusion mechanism to the case of an elevated as compared to an equilibrium concentration of vacancies [32] is based on the assumption that the rate of "solution" of pores is proportional to the vacancy concentration gradient and the number of moving vacancies. The conclusion is drawn that the more the vacancy concentration exceeds the equilibrium concentration, the higher the shrinkage rate of pores. A kinetic equation is built on these assumptions. However, Pines overlooks the fact that the original assumption included a condition making it pos-

sible for vacancies to form in the surface of pores by the diffusion mechanism. Let us remember that the main condition for the formation of vacancies in the surface of pores is the outflow of vacancies, reducing the vacancy concentration below the equilibrium concentration. If, however, the vacancy concentration exceeds the equilibrium concentration for a given capillary pressure, determined by the curvature of the particular section of the surface, then the process runs in the opposite direction – vacancies will emerge on the surface and the volume of the pore will increase.

Studies of the interdiffusion of metals with different diffusion rates have repeatedly shown the formation and growth of pores in the region of concentration of excess vacancies. All cases where vacancy sinks were insufficient for rapid reduction of the vacancy concentration to the equilibrium level or close to it resulted in the coalescence of vacancies, and the formation and growth of pores [76]. Thus, there is no basis whatever for expecting an accelerating effect of excess vacancies on reduction of the volume of pores.

3. The idea that the flow rate is proportional to the concentration of excess vacancies cannot be reconciled with the ultrahigh flow rate of a defective crystal at room temperature. As we have seen (Chapter VII), the acceleration of flow induced by an imperfect lattice is enormous, of order 10^{30} times. If it is assumed that in the equilibrium condition 1 g-atom of metal contains only one vacancy (which, generally speaking, is erroneous) then the nonequilibrium condition ensuring the observed flow rate (with the flow rate proportional to the concentration of vacancies) must correspond to a concentration of order 10^{30} vacancies per g-atom of metal, while the number of lattice sites per g-atom is equal to 6×10^{23}.

A variation of the diffusion theory developed in recent years connecting the movement of vacancies with structural elements of a crystalline body gives no explanation of the phenomena obseved in sintering of metal powders.

The Nabarro-Herring diffusion flow mechanism, connecting diffusion creep of metals with the migration of vacancies within the limits of a section of a perfect lattice limited (according to Nabarro [77]) by low-angle boundaries or (according to Herring [71]) by ordinary intercrystalline boundaries, satisfactorily describes the diffusion flow of metal at high temperatures and in several cases gives a correct quantitative estimate of the creep rate of metal in equilibrium conditions [78], but does not hold true for metals with a defective lattice. This is due to the fact that the sources and sinks of vacancies in metals with defective lattices are not only the boundaries but also imperfections within blocks or grains. The elevated vacancy concentrations created by imperfections within a block can change the sign of the concentration gradient at the boundary, which is a source of vacancies and thus must slow down and not accelerate transfer of the substance between block or grain boundaries.

However, another interpretation of the effect of imperfections within a block on the rate of diffusion flow is also possible. It can be considered that imperfections split the lattice into finer coherent sections within which there is a directional migration of vacancies, ensuring deformation of the body, on application of external forces. Let us attempt to use these concepts to calculate the order of the dimensions of blocks in a crystal capable of flow at room temperature.

According to Nabarro, the flow of metal $1/\eta$ is linked with the self-diffusion coefficient D and the size of the coherent region L by the relationship

$$\frac{1}{\eta}=\frac{D\delta^3}{kTL^2}, \qquad \text{(VIII-1)}$$

where δ is the lattice constant.

With a constant load the deformation rate is proportional to the flow. From Eq. (VIII-1) it follows that the size of blocks ensuring the different deformation rates observed for two samples of metal under different conditions must be determined by the relationship

$$\frac{L_2}{L_1}=\left(\frac{\dot{\varepsilon}_1}{\dot{\varepsilon}_2}\right)^{\frac{1}{2}}, \qquad \text{(VIII-2)}$$

where L_1 and L_2 are the block sizes, and ε_1 and ε_2 are the deformation rates of the first and second samples of metal.

In the experiments described in Chapter VII the ratio of the deformation rates of a defective and an annealed crystal was 10^{30}. It follows that with the relationship described by (VIII-1) the ratio of the sizes of coherent regions should be of order 10^{15}. This indicates that with the size of the coherent region in annealed metal $\sim 10^{-4}$ cm the size of the coherent region in a defective crystal should be of order 10^{-19} cm. Thus, the calculation by Eq. (VIII-1) gives a completely erroneous value for the coherent region of a defective crystal (let us remember that the radius of an atom is of order 10^{-8}cm). The Nabarro-Herring theory cannot explain the enormous difference in the deformation rates of crystalline particles nor the sharp difference in the densification rates during sintering of real metal powders prepared by different methods. The diffusion mechanism of creep is usually considered to act at high temperatures, near the melting point. Nevertheless, we thought it expedient to make the calculations given above, since the very high flow of defective crystals (viscous flow in outward appearance) makes it possible to assume the existence of a mechanism that does not fit into the usual framework. Such an unusual (for low temperature) mechanism would be vacancy migration with an ultrahigh concentration maintained by powerful vacancy sources. In view of the experimental data obtained, this explanation and the attempts to use other variations of the diffusion mechanism, including the formal application of Frenkel's theory to the case of nonequilibrium concentration of vacancies that was proposed by the present author [56], are clearly ineffectual.

The structural theories of creep developed in the past 20 years are based on the dislocation theory. Recent studies indicate that the dislocation mechanism can be used to explain creep even at small loads. Minute slip at stresses less than 10 g/mm^2 has been observed in aluminum single crystals [79, 80], which indicates the operation of Frank-Read sources at low stresses. In comparatively perfect copper crystals (dislocation density 10^4-10^6 cm^{-2}) dislocation movements begin at stresses as low as 4 g/mm^2, with multiplication of dislocations above 18 g/mm^2 [81]. Since the stresses induced by surface tension within the limits of a small section of a particle are of order 10^2 g/mm^2 (in particular, for a cylindrical protrusion of a copper particle 1 μ in diameter the difference in the capillary pressures is approximately 250 g/mm^2), the dislocation mechanism of flow must be regarded as possible in this case. Difficulties arise, however, in attempting to connect the dislocation mechanism of creep with the kinetics of densification and other phenomenological features of sintering.

From the formal signs the variation of the densification of active powders is outwardly similar to nonsteady creep, described by the "exhaustion" theory. In variations of this theory, Mott and Nabarro [82], Cottrell [83], and Smith [84] regard the flow of metal as the result of numerous elementary acts of slip, which are movements of individual sections of dislocations. The increase of energy required for a jump results from thermal fluctuations. It is assumed that the jump gives some constant increment of deformation, each element making just one jump.

According to these theories, the creep rate is proportional to the number of slip elements and decreases exponentially with time. Calculation of the change in flow with time during sintering of metal powders by the method given in Chapter X gives a variation that is completely different.

However, there are still more serious objections to this mechanism. The exhaustion theory is more or less justified with a small dislocation density – the creep of whiskers, for example [85]. With a high dislocation density (in deformed or rapidly grown crystals) a dense network of intersecting dislocations prevents dislocation movements and flow of the metal. Deformed metal generally undergoes little creep at room temperature or a slightly higher temperature. This was observed in a comparison of the creep rates for deformed and annealed copper at temperatures below 200°C [64]. Thus, the dislocation mechanism in the pure form cannot be used to explain the high flow rate of defective crystals at relatively low temperature.

In recent years we have seen the development of different variations of the creep theory based on a combination of the dislocation and diffusion mechanisms. At high temperature and low stress the temperature dependence of the creep rate is determined by the activation energy, which coincides with the activation energy of self-diffusion. There are also indications that the dislocation mechanism participates under these conditions (the number and distribution of dislocations change, sometimes with formation of a polygonal structure). These observations have led to the development of a creep theory based on dislocation climb (Weertman [86], Mott [87]).

The process controlling the creep rate under these conditions is the diffusion of vacancies from sources to sinks, which are sections of moving dislocations. In the absence of excess vacancies the dislocations are pinned by numerous obstacles at small loads. Circumvention of the obstacles is possible only with absorption of vacancies and climb of dislocations into neighboring interatomic planes.

This mechanism explains the laws of steady creep at small loads and was not intended for describing a process occurring at a decreasing rate. If, despite the small probability of the diffusion mechanism taking part in low-temperature creep, an attempt is made to use Weertman's equation to calculate the possible deformation rate of a defective crystal at room temperature, then analysis leads to the following conclusions.

According to Weertman, the creep rate is proportional to N/h, where N is the number of dislocations per unit volume acquiring the ability to move after climb; h is the average distance a dislocation moves with climb. In view of the fact that h should increase with N, since it would be more difficult to find a plane with no obstacle at large N, the creep rate should increase to a lesser degree than N. But even if it is assumed that the creep rate is simply proportional to N, disregarding the inhibiting effect of increasing h, the increase of the high flow rate in a defective crystal by a factor of 10^{30} is not explained by the high dislocation density, which is 10^{12}-10^{14} cm^{-2} in metals (again, we disregard the mutually inhibiting effect of dislocations at high dislocation densities, which makes the discussion above a formality and essentially unreal for this reason).

We shall not discuss other variations of the creep theory based on dislocation and diffusion mechanisms, since the analysis presented above applies equally well. The general result is that the simple dislocation mechanism is unsuitable because, under conditions ensuring flow of defective crystals at room or slightly higher temperatures, the annihilation of dislocations, to which the elimination of imperfections leads in this case, must lead to an increase and not decrease of the flow. Nor can the diffusion mechanism ensure a high flow rate at low temperatures, even at the highest possible concentration of excess vacancies. Finally, the formal use of kinetic equations derived on the basis of the assumed effect of these mechanisms (without regard to the existence of the mechanism itself) leads to the conclusion that the number of atoms in the metal is insufficient for the formation of the number of defects necessary to ensure ultrahigh flow of a defective crystal.

Thus, among the well-known physical concepts of lattice imperfections and their interaction we find no mechanism capable of explaining the ultrahigh flow of a defective crystal formed in rapid electrolytic deposition of the metal from a solution of its salts.

And there are other facts in low-temperature sintering that have not been satisfactorily explained. Active nickel and copper powders of different origin undergo noticeable shrinkage at 300-500°C, with shrinkage at a comparatively high rate in the first minutes of heating. Under these conditions the probability of the diffusion and dislocation mechanisms (as normally interpreted) taking part in densification is almost as low as in the previously described case of flow of a defective crystal at room temperature. However, there is also ground for doubting that the classical creep theories can explain the enormous change in the densification rate under ordinary sintering conditions. This follows, for example, from comparing the densification rates (reduction in volume of pores) in the beginning of low-temperature sintering and at the end of high-temperature sintering. If we assume that the temperature dependence of the densification rate is determined by the activation energy of self-diffusion, in conformity with the concepts of Nabarro and Weertman, then it is possible to calculate the densification rate from the experimentally determined rates for material sintered first at high temperature and then at low temperature.

Comparison of densification rates before and after high-temperature sintering shows a vast difference in the rates that is not linked with the theory of diffusion flow. For example, after rapid heating (in 2 min) of green compacts of active nickel powder to 600°C the momentary rate of densification was 20 cm^3/h per cm^3 of pores (i.e., 20 h^{-1}). After sintering twice at intermediate temperatures (to lower the activity of the material with relatively small densification) and three times at 1000°C the densification rate was 0.1 h^{-1}. Calculations for low-temperature sintering of material first sintered at high temperature show a densification rate of $2 \times 10^{-7}\,h^{-1}$ at 600°C. The ratio of the rates before and after high-temperature sintering amount to 10^8.* This means that if Nabarro's mechanism controls the flow then the ratio of block sizes before and after high-temperature sintering should be of order 10^{-4}, which is practically unattainable (if the size of the coherent regions in annealed material were of order 1 μ then, according to Nabarro's theory, in an active material the size of blocks ensuring the observed ratio of rates should be of order 1 Å, i.e., less than the size of the atom).

From the totality of observations one can conclude that densification in low-temperature sintering depends on a special mechanism whose physical nature is still unknown. Also, there is the question of whether this mechanism can operate at higher temperatures and whether there is a "transition region" in which the low-temperature mechanism coexists with the better-known mechanism (or mechanisms ?) of high-temperature creep. As will be shown below (Chapter XIII), attempts to observe special densification laws for high-temperature sintering that differ from densification in low-temperature sintering have failed. One gets the impression that the densification mechanism is the same in a wide temperature range.

In connection with the difficulties that occur in using modern creep theories to explain the laws of densification, phenomenological generalizations of the kinetics of densification become important as a necessary stage in formulating a theory of sintering. At the same time, attempts to arrive at phenomenological generalizations should not be opposed to efforts to formulate a physical theory where (in the present condition of the theory) a qualitative but well-founded explanation of the special features of the kinetics would be an important step in developing a sintering theory.

*The activation energy of flow found with a rapid increase of the temperature between two periods of isothermal sintering is larger than the activation energy of self-diffusion of nickel and therefore the actual ratio of the rates exceeds 10^8.

Chapter IX

Quantitative Estimate of the Effect of the Geometric Factor

The characteristics of densification of porous crystalline substances previously considered reflect mainly the effect on the flow rate of lattice defects in crystalline substances (substructural factor). The densification rate also varies with the distribution and magnitude of capillary pressures associated with the shape, size, and relative concentration of pores. The effect of the geometric factor (geometric characteristic of the pores) is reflected in the variation of the densification rate with the dispersity of the powder.

The goal of phenomenological analysis is a quantitative estimate of the effect of both factors, including a mathematical description of the change in the effect with time.

Empirical equation (III-5), describing the reduction of pore volume during isothermal sintering, reflects the simultaneous effects of the substructural and geometric factors. The separation of the effects of the two factors is based on the assumption that sintering of a crystalline substance differs from sintering of an amorphous substance only in the change of flow with time in the crystalline substance because of the reduction of the concentration of defects, while the flow of the amorphous substance is constant at a given temperature. Therefore the effect of the geometric factor can be determined by means of establishing the change in the volume of pores with time for a porous amorphous body.

Another method is also possible. Determining the effect of the geometric factor amounts essentially to establishing the form of relationship $dv/d\tau = f(\tau)$ or $dv/d\tau = \varphi(v)$ for the case of sintering with a constant concentration of defects. The form of the first relationship is difficult to establish experimentally for crystalline bodies sintered under ordinary conditions. Concentrations of defects constant in time can be created only by special methods – for example, by irradiation with fast neutrons of constant power (the experimental data obtained under these conditions will be considered below). It is simpler to determine the form of the second relationship $dv/d\tau = \varphi(v)$, which can be determined indirectly on the basis of the experimental data characterizing the variation of $dv/d\tau$ with v for a given sintering time.

During sintering under the same conditions of porous crystalline bodies with different initial porosities prepared from the same powder the concentration of defects changes with time in the same way, and at any time during sintering the concentration of defects is the same in all bodies. The variation of the rate of reduction in volume of pores with the volume of pores can be determined by comparing these values for bodies with different initial porosities sintered under identical conditions. As was shown earlier (Chapter I), the relative reduction in volume of pores is the same in such bodies. This means that at any given time during sintering the rate of reduction in volume of pores is proportional to the volume of pores. This conclusion also holds true for bodies with a difference in porosity sintered at the same time. However, it can be assumed that the same relationship, i.e., proportionality of $dv/d\tau$ and v, would also be observed after sintering of the same porous body for different times if the flow does not change with time. Generally speaking, there should be a difference between the characteristics of the pores in bodies sintered at the same time and in a body sintered for different times – in the latter case the pore surface should be smoothed out more after sintering for a long time. However, smoothing of the surface or changes in the shape of pores obviously do not in themselves make substantial changes in the kinetics of densification (see Chapter VI). This follows, for example, from the constant relative reduction in volume of pores: With an increase of compacting pressure the shape of the pores changes substantially due to deformation of the particles, but this is not reflected in the reduction in volume of pores during sintering, which remains constant for bodies with different densities and different shapes of pores.

If the assumption of identical effects of changes in the volume of pores during compacting and sintering on the rate of reduction in volume of pores holds true, then from the constant relative reduction in volume of pores with different initial pore densities we have a proportionality of the rates of reduction in volume of pores that is also a mathematical expression of the effect of the geometric factor on reduction in volume of pores during sintering, i.e.,

$$\frac{dv}{d\tau} = -\xi v, \qquad \text{(IX-1)}$$

where ξ is a coefficient with a value depending on the size and geometric characteristics of the original particles and the current value of the viscosity of the substance. For crystalline bodies ξ changes with time due to the variation of flow with the concentration of defects. Since the difference in the kinetics of the densification of crystalline and amorphous bodies consists only in the fact that the viscosity changes for the first and is constant for the second, Eq. (IX-1) must also be valid for sintering of amorphous bodies. This can be checked by comparing the reduction in volume of pores during sintering of glass powder with relationship $v_s/v_p = f(\tau)$ from Eq. (IX-1). Sintering was conducted in the dilatometric apparatus described above. At the selected sintering temperatures the viscosity of glass was fairly high, which made it possible to observe densification over a period of several hours. The variation of the volume of pores with time for glass powder was investigated in several temperature ranges (Fig. 27a). The temperature was held constant in each experiment within limits of $\pm 3°C$. The difference in the temperatures of the experiments was only 10°, since a small increase in temperature substantially reduces the viscosity of glass.

Integration of Eq. (IX-1) gives a kinetic equation linking the volume of pores with sintering time:

$$\ln \frac{v}{v_{in}} = -\xi\tau. \qquad \text{(IX-2)}$$

It follows from Eq. (IX-2) that the variation of log (v_s/v_p) with τ for the experimental data should be linear. This relationship for glass powder is shown in Fig. 27b. At the beginning of

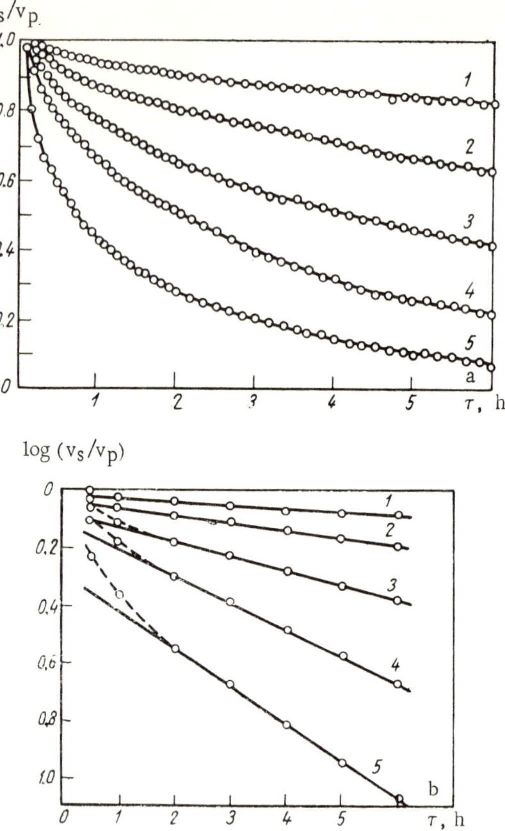

Fig. 27. Plots of v_s/v_p vs. τ (a) and log (v_s/v_p) vs. τ (b) for glass powder at different sintering temperatures. 1) 580°C; 2) 590°C; 3) 600°C; 4) 610°C; 5) 620°C. The dashed lines show the deviation of the experimental data from a linear relationship.

sintering the experimental data deviate from a straight line, but fall approximately on a straight line after sintering for a certain time. The deviation from a straight line is probably due to irregular densification in the initial period of sintering. At the beginning of sintering it is possible for the number of contacts between particles to change, along with rapid shrinkage of the smallest pores, filling of small fissures, etc. Regular reduction in volume of pores in conformity with Eq. (IX-2) becomes possible only after the geometric characteristic of the pores becomes stable (stable contacts and connected channels).

These same phenomena must also occur during sintering of crystalline powders. An indirect confirmation of this can be seen in the fact that the relative reduction in volume of pores is strictly constant mainly for powders with particles of nearly the same size and shape. For powders with particles of complex shape that differ in size v_s/v_p frequently deviates from constant. In most cases these deviations are due to processes developing in the beginning of sintering when the structure of the pores is stabilized; this is indicated by the fact that during resintering at the same temperature or a higher temperature v_s/v_p is constant even though it is not constant during presintering. As an example, results are given in Table 24 for sintering of copper powder obtained by reduction of oxides in hydrogen at a relatively low temperature (400°C). The particles of copper powder are irregular conglomerates of fine grains. As in the case of sintering once, after sintering twice v_s/v_p is not constant, but during resintering the relative reduction in volume of pores at a different initial density is practically the same. It can be assumed that the additional contacts occurring with an increase of compacting pressure as well as the formation of new contacts during sintering disrupt the regular densification that occurs for bodies with stable interconnected pores. As already noted, other processes that disrupt the regular course of densification at the beginning of sintering are also possible –

TABLE 24. $v_s/v_p = f(d_p)$ for Single and Double Sintering of Copper Powder*

Green density, g/cm³	Value of v_s/v_p after sintering		
	Single sintering, 680°C, 30 min	Double sintering, 680 and 850°C, 30 min	Second sintering, 850°C, 30 min
2.65	0.526	0.320	0.608
3.02	0.512	0.313	0.611
3.42	0.457	0.280	0.612
4.10	0.453	0.274	0.605
4.44	0.421	0.253	0.602
5.03	0.420	0.260	0.620
5.72†	0.509	0.366	0.720†
6.35†	0.680	0.542	0.806†

*Powder obtained by reducing lumps of copper oxide in hydrogen at 400°C.
†With a green density over 5.50 g/cm³ the value of v_s/v_p increases due to the development of the expansion process.

for example, breaking and shifting of contacts under the influence of the elastic-plastic aftereffect.

The phenomena disrupting the regular course of densification are not reflected, however, in the reduction in volume of pores during isothermal sintering because these phenomena cease before the beginning of isothermal sintering. Rapid formation of stable channels in crystalline bodies is ensured by the very high flow of the substance at the beginning of sintering. The formation of stable pores requires some initial densification, which in sintering of crystalline bodies is reached in the first minutes of sintering, during heating to the isothermal sintering temperature. The high densification at the beginning of sintering of crystalline bodies can be judged from the following data: The densification reached after 8 min of heating to a given temperature sometimes exceeds the densification occurring in the following 50 h of isothermal sintering. In all our experiments densification in the first 8 min exceeded one-third of the total densification after 50 h of sintering (see Table 7, Chapter III).

The course of densification in porous amorphous bodies differs substantially from densification of crystalline bodies. Because of the constant flow the rate of reduction in volume of pores does not change with time. The densification during the time of heating to the given sintering temperature is exceedingly small (usually less than 3% of the total densification). At temperatures suitable for investigation, i.e., ensuring slow sintering, the formation of stable channels continues over a substantial portion of the isothermal sintering period. It must be assumed that the deviation of log (v_s/v_p) vs. τ from linear in the beginning of sintering of glass powder coincides with the formation of stable channels, upon completion of which the course of densification conforms rather closely to Eq. (IX-2).

Thus, except for the period in which stable channels are formed the experiment confirmed the possibility of describing the reduction in volume of pores of an amorphous body with Eqs. (IX-1) or (IX-2).

The densification of a porous amorphous body was analyzed by Geguzin [88], who, on the basis of Frenkel's theoretical assumptions [14], derived a kinetic equation for the reduction in the volume of pores in a body with cylindrical channel-like pores. Comparison with the experimental data [55] showed that this equation satisfactorily describes the shrinkage of a porous body of glass powder. The theoretical and empirical equations give similar results. From comparison of the two it follows that empirical equation (IX-2) is not in conflict with the description of the process based on Frenkel's theoretical assumptions. Also, Geguzin's theoretical analysis does not take into account the numerous intersections of channels and the presence of sections of surface with compound curvature in the area of intersections. Thus, as will be shown below, it does not fully agree with the experimental data.

After modification in connection with the description of the reduction in volume of pores, Geguzin's equation [88] takes the form

$$\frac{v}{v_{in}} = (1 - K\tau)^3, \qquad \text{(IX-3)}$$

in which

$$K = \frac{\sigma}{\eta \cdot r_0},$$

where σ is surface tension, η is viscosity, and r_0 is the initial pore radius.

Comparison of the curves of Eqs. (IX-2) and (IX-3) with the experimental data (any of the curves in Fig. 27a) indicates that after the interconnected pores are formed, i.e., disregarding the initial sections of the curves, empirical equation (IX-2) more accurately describes the reduction in volume of pores than Eq. (IX-3). Figure 28 shows the variation of log (v_s/v_p) with τ and $(v_s/v_p)^{1/3}$ with τ plotted from the same experimental data given in Fig. 27a. When Eq. (IX-2) or (IX-3) fits the experimental data the variation should be linear. The experimental points are close to linear in coordinates of log (v_s/v_p) vs. τ, which indicates that Eq. (IX-2) describes the experimental data more accurately. It should be noted that the curves for both equations do not coincide with the experimental curves in the beginning of sintering, i.e., in the period when interconnected pores are formed. Equation (IX-3) has no advantage over (IX-2) in this connection.

There is no doubt that Eq. (IX-2) gives only an approximate description of the reduction in volume of pores with constant flow.

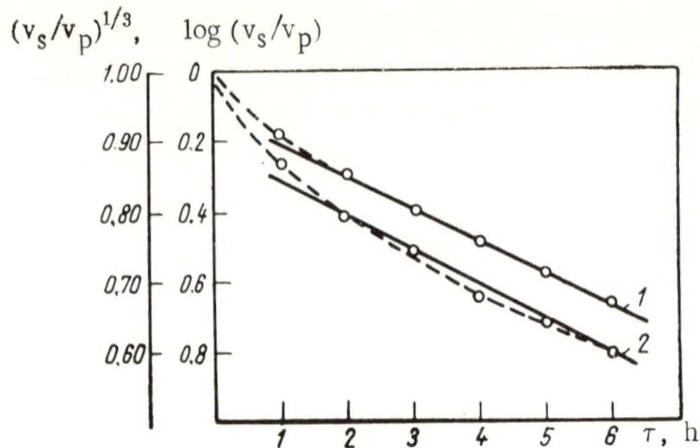

Fig. 28. log (v_s/v_p) vs. τ (1) and $(v_s/v_p)^{1/3}$ vs. τ (2) plotted from experiments with isothermal sintering of glass powder. The experimental data are shown by dashed lines and the variations described by Eqs. (IX-2) and (IX-3) by solid lines 1 and 2 respectively.

The possible error would hardly be large, since the effect of the geometric factor on the course of densification is negligible in comparison with the effect of the substructural factor. Therefore the error due to inexact calculation of the effect of the geometric factor is hardly reflected in the values calculated by the general kinetic equation accounting for the effects of both the geometric and substructural factors.

In our calculations we shall use the empirical equation (IX-2) for the following reasons. During the period when stable pores are established Eq. (IX-2) is as accurate in describing the densification of bodies with a constant viscosity as Eq. (IX-3). Equation (IX-3) does not conform with one of the main rules of densification – constant reduction in volume of pores (the initial pore radius enters into the term for coefficient K). The form of Eq. (IX-3) complicates its use for analysis of the kinetics of densification and the rules expressed by (III-5), while Eq. (IX-2) is easily used for this purpose.

These same reasons compel renunciation of the equation introduced in [5] to describe densification of an ideally viscous porous body with closed pores for which v_s/v_p is not constant at different d_p. In addition, the equation does not match the structure of the pores we observed (interconnected channels and not isolated pores).

Kingery's calculations [89] give the appearance of a theoretical basis of the constant relative reduction in volume of pores and the laws expressed by Eq. (IX-2). However, the derivation contains an error in logic. After replacement of the number of pores in Mackenzie's and Shuttleworth's equation with a value expressing the total volume of pores and the radius of pores, the radius is later taken as a constant value, which leads to an incorrect conclusion concerning the reduction in volume of pores due to the reduction of their number with retention of a constant radius. When the assumption is corrected (n constant, r variable) the replacement does not lead to simplification of the Mackenzie-Shuttleworth equation, and the equation obtained does not conform with constant v_s/v_p.

Experimental data have been obtained recently that directly confirm the suitability of Eqs. (IX-2) and (IX-3) to describe the densification of crystalline bodies with constant (not changing in time) flow. It was found in [90] that when porous graphite obtained by deposition of carbon from the gaseous phase is bombarded with fast neutrons of constant power the densification is uniform at temperatures of 500-1000°C. It was shown that the rate of reduction in volume of pores decreases proportionately to the change in volume, and the extent of reduction in volume of pores is independent of the initial porosity. In other words, it was shown that with continuous irradiation the kinetics of densification obeys the rule described by Eq. (IX-1). It can be assumed that constant power of the neutron flux in a crystalline body leads to a quasiequilibrium (constant) concentration of defects, determined by the equality of defects formed and defects disappearing per unit time. Under these conditions the course of densification reflects the effect of the geometric factor alone, since the concentration of defects is constant. Thus, these studies again confirm the possibility of describing the effect of the geometric factor during sintering of crystalline bodies by means of Eq. (IX-1).

In applying (IX-1) to sintering of crystalline bodies it should be kept in mind that the effect of the geometric factor is determined by the constant component of coefficient ξ, varying only with the geometric characteristic of the particles of the original powder, which determines the geometry of the pores. Its other component, characterizing the flow of the substance in the presence of concentrations of defects, changes with time. Let us call the constant

β and the variable component γ. After replacing ξ in Eq. (IX-1) with the product $\gamma\beta$, we have

$$\frac{dv}{d\tau} = -\gamma\beta v. \qquad \text{(IX-4)}$$

The effect of the geometric factor on the rate of reduction in volume of pores is determined by the product βv, which takes into account both the effect of the original geometric parameters (size and shape of particles of the original powder) and the change in the geometric factor with decreasing v. The value of the flow $[\gamma = (1/\eta) P^{-1}]$. It changes with time and, being dependent on the concentration of defects, characterizes the effect of the substructural factor.

To complete the analysis of the empirical equation describing the reduction in volume of pores it is necessary to determine the variation of γ with time. This variation will characterize the effect of the substructural factor on the densification process.

Chapter X

Quantitative Estimate of the Effect of the Substructural Factor

The quantitative characteristic of the change of flow with time during sintering due to the influence of the substructural factor can be found by means of comparing the general densification law expressed by Eq. (III-14), reflecting the combined effect of the geometric and substructural factors on the densification process, with Eq. (IX-4), characterizing the effect of the geometric factor alone. This method made it possible to propose a mathematical description of the change of flow with time about 15 years ago [56].

However, it is of interest not only to determine the change of flow with time but also to connect the laws governing flow with elementary processes that can be described by equations including the kinetic characteristics of reactions (activation energy and order of reaction). The goal was to determine whether it is possible to establish the form of the variation of flow with concentration of defects and the quantitative characteristic of the relationship, which is determined by the change in the concentration of defects with time.

Let us proceed with the first variation of the analysis on the assumption that the flow rate is directly proportional to the concentration of defects. As is well known, formal (structureless) theories of creep are based on a similar assumption that the creep rate is proportional to the number of "activation complexes" or "elementary slip zones" (Kauzmann [91], Mott and Nabarro [82], and others). It cannot be considered that this analogy exhausts the basic premises, but it indicates that it is reasonable. Later we shall consider other possibilities.

To determine the form of the kinetic equation describing the change in the concentration of defects with time it would be logical to revert to studies of the kinetics of recovery. However, we find no successful analogies (at least analogies that would indicate directly the most probable form of the kinetic equation describing the elimination of defects during sintering).

Studies of the recovery of the mechanical and physical properties of deformed metals have produced more than one theory of recovery. The kinetics of recovery have proven to differ for different metals, temperature ranges, deformation conditions, or formation of lattice defects.

The first attempts to describe recovery as the consequence of the interaction of dislocations [92, 93] were of a formal nature, since the properties of dislocations were not in fact used, and the same conclusions can be reached equally well by use of the structureless concepts of the "elementary act of softening" and "concentration of defects." Kuhlman [94-96] considered recovery to be the result of the interaction and annihilation of dislocations. It was found that the use of Becker's concept [97] of the variation of the activation energy of dislocation movements with shearing stress leads to a kinetic equation of recovery in which the activation energy proves to be dependent on the degree of recovery attained. Cottrell and Aytekin [98] later reached a similar conclusion. They found that when the variation of the activation energy with shearing stress is linear (the latter increases with dislocation density) the variation of recovery with time takes the form

$$M = B - A \ln \tau, \tag{X-1}$$

where M is the level of the property depending on dislocation density, and A and B are constants.

This dependence is actually observed in some cases of recovery of deformed metals – nickel [99], aluminum single crystals [100], zinc single crystals [98]. However, large numbers of experimental data do not follow a logarithmic relationship. Several authors [101, 102] indicate that using the derivation of Eq. (X-1) on the assumption of a linear variation of the activation energy with an increase in the level of the property due to strain hardening and impairment of the property during recovery is not fully justified.

Studies of the liberation of energy after strain hardening or other means of creating lattice defects have shown a complex temperature dependence of this process, often with several peaks, indicating that intensive elimination of defects can occur in different temperature ranges separated by regions where the development of the process is relatively slow. Comparison of the maximum liberation of energy with changes in the properties indicate that the region of intensive liberation of energy may coincide with changes in different properties.

Various methods of creating defects (deformation at different temperatures, radiation damage, quenching from high temperatures, rapid growth of crystals at low temperatures) lead to the formation of crystals with different ratios of defects of different types. The different types of defects, or more often combinations of different defects, are responsible for the change in properties during annealing. Also, the most intensive changes in different properties frequently do not occur at the same temperatures. All this indicates the complex nature of recovery and the far from identical character of the variation of different properties with lattice defects (for example, see the review article by Bever [101]).

Changes in the properties of crystalline bodies can be induced by the simultaneous influence of different types of lattice defects, the elimination of which depends on their activation energies, and possibly also different time dependences. It remains unclear whether the elimination of a given type of defect with a constant activation energy or the activation energy itself depends on the concentration or on the difference in the structure of individual defects (in a given type of defect). The latter case conforms with the assumption of a "spectrum" of activation energies of defects of a given type.

Finding a distribution function with a complex spectrum of activation energies is extremely complicated. Vand [103] attempted to solve this problem in general form, proposing a method of calculating the distribution of activation energies of the elimination of defects from the change in the electrical resistivity during isothermal annealing and also annealing with the temperature increasing at a constant rate. However, this method proved to be difficult to use for other types of recovery. Several authors [104] have questioned the validity of Vand's method. Another method of calculating the distribution function was proposed by Primak [105]. Primak's method is very complicated, and it is also complicated to use the activation energy distribution to describe recovery at different temperatures.

Special forms of the distribution function of the activation energy [106], including standard Gaussian distribution [107], have been used to analyze several processes. The temperature dependences derived in these attempts do not match the temperature dependence of flow observed in sintering, and the methods proposed by the authors cannot be used to analyze the temperature dependence of densification for porous bodies.

Despite the complex nature of recovery, relatively simple time dependences are often observed that are described by equations of chemical kinetics. The temperature dependence of recovery is often determined by a constant value of activation energy. There have been many investigations in which such dependences were found, and we shall limit ourselves to a few typical examples. A study of recovery in deformed copper was made by determining the thermo-emf of deformed and annealed samples coupled together. It was found that the process is described by an equation with a single value of the activation energy (~22,000 cal/g-atom) at temperatures of 100-300°C [108]. From the change in the electrical resistivity of the deformed copper a constant value of the activation energy of recovery was found, equal to 28,000 cal/g-atom [109]. Recovery was investigated in the temperature range of 100-250°C. The recovery of deformed copper in approximately the same temperature range was investigated by measuring the intensity of x-ray diffraction lines [110]. A constant value of the activation energy was found for spectrally pure copper (22,400 cal/g-atom) and commercial copper (29,900 cal/g-atom). It is interesting to note that the change in the properties due to the recrystallization process investigated in a wider temperature range is also frequently characterized by a constant activation energy of the process determining the change in the properties.

Similar observations were made in studies of recovery in other metals. With small deformation within the limits of easy slip a constant activation energy of recovery of mechanical properties was found for a deformed single crystal of zinc [102]. The activation energy was independent of previous deformation. With considerable deformation the variation of the activation energy of recovery with prior plastic deformation has been observed in several cases. Thus, a study of recovery in deformed aluminum by measurements of the electrical resistivity and hardness showed an activation energy of 32,100 cal/g-atom with 20% deformation and 22,500 cal/g-atom with 80% deformation [111, 112]. However, within the limits of some degree of deformation the activation energy is constant and does not change with an increase in the extent of recovery. A reduction of the activation energy of recovery was observed in commercial iron [113] with an increase of deformation. The activation energy reached a minimal value at large deformation, and then remained unchanged with further increase of deformation. A constant value of the activation energy of recovery, independent of the deformation, was found for commercial iron with 65% deformation or more [114].

As already mentioned, a change in the value of the activation energy with increasing extent of recovery is often noted. In the recovery of metal deformed at low temperatures (4-10°K) a constant value of the activation energy was found in several temperature ranges, and a variable value at other temperatures [115]. It can be assumed that at least in some cases the change in the activation energy with the temperature or extent of recovery is due to the combination of several recovery mechanisms.

Based on the results of recovery investigations, a phenomenological analysis of a phenomenon close to recovery (reduction of flow due to elimination of defects) must begin with the assumption that the activation energy is constant, and more complex assumptions should be made only in the case where the experimental data indicate that the activation energy is not constant.

Since "structural" theories of recovery based on an exact physical characteristic of defects are still far from a universal kinetic law confirmed by experiment, to derive an equation describing the change in flow during sintering one can use assumptions and hypotheses of the structureless (phenomenological) theory of recovery, which has been more successfully used to describe the time dependence of recovery. In many studies of recovery it has been shown that the process can be described approximately by equations of chemical kinetics. Along with attempts at a more rigorous interpretation of recovery, work of this nature has been appearing for 20 years. Not long ago an attempt was made to describe recovery completely on this basis, with the use of concepts from the theory of absolute reaction rates and the concept of an activation complex – a section of the lattice capable of rebuilding to the "state of conversion thermodynamic instability" [116].

Authors using concepts of chemical kinetics pose the problem of determining the "order of reaction," i.e., the exponential factor, in which the concentration of defects or an increase of a property due to a high concentration of defects is significant. The following equations is used for a phenomenological description of recovery:

$$\frac{dx}{d\tau} = -\vartheta x^n \exp\left(-\frac{E}{RT}\right), \tag{X-2}$$

where x is the current numerical value of the increase in the property (difference in the level of the property in a sample being annealed and in a completely annealed sample). It is assumed that the value of x is proportional to the concentration of defects. Exponent n determines the reaction order. By analogy with chemical kinetics, it is assumed that the value of n matches the number of interacting activated complexes or defective sections of the lattice.

Different values of n from 1 [108, 110] to 4 [109] have been obtained in studies in which the course of recovery in different metals with various means of creating defects has been compared by means of Eq. (X-2). Although recovery is satisfactorily described by Eq. (X-2) with constant values of n in many cases, no common kinetic laws have been established. In individual cases recovery follows second-order kinetics rather closely, which seems to be the best fitted to the interaction of defects. Second-order kinetics was observed in recovery of the electrical resistivity of copper near room temperature after neutron bombardment at 80°K and lower [117, 118] and in recovery of the electrical resistivity of a copper film formed by condensation on a cooled substrate [119]. Near second-order kinetics was observed in the recovery of the electrical resistivity of gold quenched from 500 to 950°C, with recovery observed at 10-15°C [120-121].

Second-order kinetics is not observed in the recovery of the properties of deformed metal. In subsequent annealing of samples with a defective lattice created by mechanical means second-order kinetics was not observed in all cases, and for the most part only in a limited temperature range. It is very possible that fresh interactions of defects corresponding to a second-order reaction are often encountered, but the corresponding kinetic law is not manifest during annealing because of the concurrent development of different recovery mechanisms.

The change in flow during sintering can evidently be regarded as a manifestation of a certain type of recovery. Therefore Eq. (X-2) seems suitable for a phenomenological description. The problem will be to determine the order of reaction by comparing kinetic equations having different n with the experimental data. The method usually used to determine the order of reaction [109, 122] amounts to the following. Substituting in Eq. (X-2) values of n equal to 1, 2, and 3, we obtain, after integration, three equations that correspond to first-, second-, and third-order kinetics:

first order

$$\ln x = \ln x_{in} - \delta\tau; \qquad \text{(X-3)}$$

second order

$$x = x_{in}(1 + \delta x_{in}\tau)^{-1}; \qquad \text{(X-4)}$$

third order

$$x = x_{in}(1 + 2\delta x_{in}^2\tau)^{-\frac{1}{2}}. \qquad \text{(X-5)}$$

Here, $\delta = \vartheta \cdot \exp(-E/RT)$ from Eq. (X-2); x is the current value of the difference in the levels of the property in the sample during annealing and in the fully annealed sample; x_{in} is the same for the beginning of annealing.

The experimental data can be conveniently compared with these equations by plotting $\log x = f_1(\tau)$ for (X-3), $1/x = f_2(\tau)$ for (X-4), and $1/x^2 = f_3(\tau)$ for (X-5). On graphs plotted in these coordinates the variation should be linear.

To determine the relative flow one should use the experimentally established rate of reduction in volume of pores (determined by constructing tangents and measuring their slopes on plots of v_s/v_p vs. τ). The relative flow after different isothermal sintering times can be found from these values if the effect of the changing geometry of pores on the densification rate is taken into account and disregarded.

As was established earlier (Chapter IX), the effect of the geometric factor amounts to the fact that the multiplier characterizing the effect at a certain densification rate decreases proportionally with the volume of pores v.

In the absence of change in the volume of defects

$$\frac{dv}{d\tau} = -\xi v.$$

In this case ξ is constant. With changing concentrations of defects this same relationship holds true, but ξ changes with time, reflecting the change in flow. Thus, in order to find the value of ξ at the current value of the volume of pores, equal to v_i, it is necessary that the experimentally determined rate of the reduction in volume of pores be divided by the volume of pores

$$\xi_i = \left(\frac{dv}{d\tau}\right)_i \cdot \frac{1}{v_i}, \tag{X-6}$$

where ξ_i is proportional to the flow ("relative flow"), and $(dv/d\tau)_i$ and v_i are the rate of reduction and the volume of pores at a given isothermal sintering time τ_i.

The change in flow was matched with one of the kinetic equations [(X-3), (X-4), or (X-5)] by using the values of ξ found by the method given above for one of the typical functions $v_s/v_p = f(\tau)$ (sintering of copper; see Table 7). These values were used to plot $\log \xi$ vs. τ, and $1/\xi$ vs. τ, and $1/\xi^2$ vs. τ (Fig. 29). A linear relationship was obtained for $1/\xi = f(\tau)$, which indicates that the change in flow conforms with second-order kinetics.

A similar analysis could be made for many experimentally determined $v_s/v_p = f(\tau)$, although it is not necessary. It is easy to show that from the time dependence of flow conforming to second-order kinetics comes a change in the volume of pores with time established by Eq. (III-5). The fact that the reduction in volume of pores matches this equation is already a confirmation that the time dependence of flow obeys second-order kinetics. Additional mathematical substantiation of this conclusion will be presented in the following chapter.

Since Eq. (III-5) accurately describes the change in the volume of pores under different sintering conditions, it can be stated that the change in flow obeys second-order kinetics for different powders sintered in a wide range of temperatures.

In view of the lack of a single law of recovery for different properties and the possible overlap of several recovery mechanisms, this conclusion is somewhat unexpected. Even though the explanations of this rule may cause doubts, the existence of the rule is established fact. When the fitness of Eq. (III-5) is checked against the results from many experiments there is not a single case of sintering for 3-4 h where the change in the volume of pores deviates substantially from the course described by this equation (see Chapter III).

The conformity of the change of flow with second-order kinetics makes it possible to describe the change of flow with time by an equation similar to (X-4):

$$\gamma = \gamma_{in}(1 + \delta\gamma_{in}\tau)^{-1}. \qquad \text{(X-7)}$$

Second-order kinetics for the time dependence of flow during sintering of metal powders was first proposed by the author in

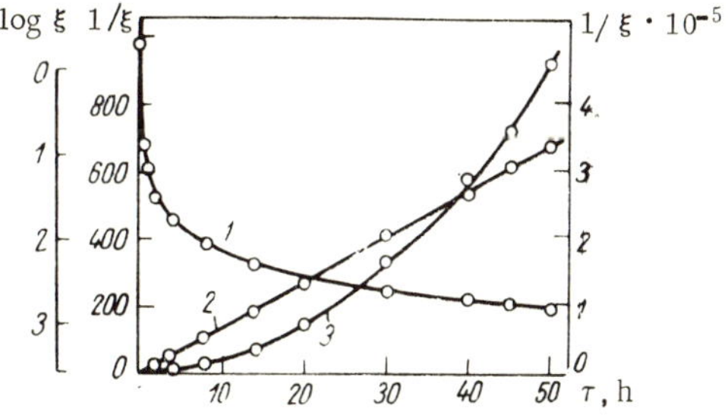

Fig. 29. log ξ vs. τ (1), 1/ξ vs. τ (2), and $1/\xi^2$ vs. τ(3) plotted from results of experiments with isothermal sintering of copper powder; value of ξ determined by Eq. (X-6).

1955 [56]. In recent years it has been confirmed by several investigators.

From Eq. (X-7) comes a linear time dependence of flow:

$$\eta = \frac{1}{\gamma} = \frac{1}{\gamma_{in}}(1 + \delta\gamma_{in}\tau)$$

or

$$\eta = \eta_{in} + \delta\tau. \qquad \text{(X-8)}$$

The linear variation of the flow with sintering time was found by Skorokhod [123]. The inaccurate estimate of the effect of the changing geometry of the pores did not prevent an accurate estimate of the change of flow with time, which is due to the fact that the change of flow has a far larger effect on the rate of reduction in the volume of pores than the change in the geometry of the pores, as previously noted. The linear variation of viscosity with sintering time was also noted by Samsonov [37] and Koval'chenko [124].

Since the linear variation of viscosity is due to the fact that that change of flow obeys second-order kinetics, these studies confirm the second-order kinetics for the time dependence of flow. Other explanations of the linear variation of viscosity have not been convincing [37, 123, 124]. The deviation of the time dependence of viscosity from the diffusion flow equation [77] and the time dependence of the increase in mosaic block size [32] proposed in [123] lead to an equation by which the increase of viscosity is proportional to σ/r after some constant time (σ is the surface tension of the block boundaries; r is the atomic radius). This ratio changes little with temperature, which contradicts the substantially lower "deceleration" of densification with increasing temperature [this phenomenon is manifest in the regular reduction of constant m in Eq. (III-5) with increasing temperature]. Thus, Skorokhod's interpretation of deceleration [123] disregards the principal feature of the temperature dependence of densification.

Several authors [37, 125] have attempted to connect the change of the densification rate with grain growth, which also leads to a linear variation of viscosity with time.

The low probability of such a dependence for sintering of metal powders was shown in Chapter VIII. Let us remember that in many cases grain growth differs from $L \sim \tau^{1/2}$, which could ensure the change of flow in conformity with second-order kinetics. The generalized equation of grain growth kinetics [126, 127] has the form

$$L^2 - L_0^2 = K\tau^n,$$

where L_0 is the original size, L is the size after time τ, and K and n are constants.

This equation matches second-order kinetics only in the particular case where $L \gg L_0$ and n = 1. This condition is not usually observed in sintering of metal powders. Thus, Evans and Ashall [128] found that grain growth during sintering of nickel powder is described by $L^2 - L_0^2 = K\tau^{0.93}$, which differs substantially from $L \sim \tau^{1/2}$. Coble [31] found $L \sim \tau^{1/3}$ in sintering of copper powder, which does not conform with second-order kinetics. In addition to observations on grain growth and densification, they speak of the fact that although both processes usually develop together, the maximum deceleration (decrease of the rate) of densification does not coincide with the period of highest grain growth. Many authors have noted the effect of the number and shape of pores in grain growth. Intensive grain growth frequently develops after notable reduction of the number and simplification of the shape of pores [128-130], while densification slows down at the beginning of isothermal sintering, when grain growth is still not very intensive.

Finally, the greatest possible change in block size, and particularly the grain size, cannot be responsible for the enormous difference in the densification rate (as was shown in Chapter VIII).

All these complications are completely removed if we proceed on the assumption that the linear variation of viscosity with time is a consequence of the fact that the elimination of defects obeys second-order kinetics. With this interpretation the temperature dependence of densification acquires a phenomenological explanation permitting a quantitative estimate of the dependence.

Thus, the totality of our studies and those made by other investigators leads to the conclusion that for a wide range of sintering temperatures and for quite different materials (including metals, carbides, and borides) the effect of the concentration of lattice defects is manifest in a change of the flow with time corresponding to second-order kinetics. As we shall see, describing the elimination of defects responsible for the increase in flow of a crystalline substance by means of a kinetic equation of a second-order reaction with a constant activation energy permits a fairly complete explanation of many phenomenological characteristics of the densification of metal powders during sintering.

Chapter XI

Phenomenological Theory of Sintering

Mathematical formulations accounting for the effect of the geometric and substructural factors reflect the kinetics of elementary processes determining the course of densification. From these data it is possible to work toward a more complete description of the densification process, including its temperature dependence.

The flow of a crystalline substance with lattice defects is a thermally activated process. However, the temperature dependence of flow is complicated by the fact that an increase of temperature has a dual effect – the flow is accelerated at a given concentration of defects, along with acceleration of the elimination of defects, so that the concentration of defects decreases with time. Therefore, the flow activation energy can be used to describe the temperature dependence only with instantaneous transition from one temperature to another. To describe the course of densification (change in volume of pores) with time at different temperature it is necessary to determine the temperature dependence of a second elementary process – elimination of defects. The kinetic laws of elementary processes (reduction in volume of pores with a constant concentration of defects) make it possible to do this. We shall proceed on the assumption that the temperature dependence of the elementary processes is determined by activation energies that are different but constant for each process.

The relationship between the rate of reduction in volume of pores and the flow is expressed by Eq. (IX-4):

$$\frac{dv}{d\tau} = -\gamma\beta v,$$

where $\gamma = 1/\eta$, and β is a coefficient depending on the initial geometric characteristic of the pores. To derive a differential equation containing only v and τ as variables it is necessary to determine the variation of γ with time.

Further analysis requires quantitative calculation of the change in the concentration of defects, and therefore N, characterizing the concentration of defects, must be introduced into the mathematical formulation. We shall not be concerned with the physical meaning or dimensions of this value. The value N can be characterized both as the absolute and relative concentration of defects, and the form of the mathematical formulation including this value will not depend on the units in which it is measured. The final equation will contain N_{in} (the initial concentration of defects) together with a constant multiplier (coefficient of proportionality), which can be selected for a value of N at any scale.

By analogy with phenomenological theories of recovery and creep we assume that flow is proportional to the concentration of defects. Considering that flow with an unchanging concentration of defects is a thermally activated process, we can write

$$\gamma = \vartheta \cdot \exp\left(-\frac{E_b}{RT}\right) N, \tag{XI-1}$$

where ϑ is the coefficient of proportionality accounting for the dimensions of N, and E_b is the activation energy of flow of a given type (the mechanism of flow associated with lattice defects). Substituting (XI-1) into (IX-4) and replacing $\vartheta\beta = b$, we obtain

$$\frac{dv}{d\tau} = -bN \cdot \exp\left(-\frac{E_b}{RT}\right) v. \tag{XI-2}$$

The flow, and consequently the concentration of defects, changes with time in conformity with second-order kinetics (see Chapter X). This means that with a constant activation energy E_a the kinetics of the elimination of defects is described by

$$\frac{dN}{d\tau} = -a \cdot \exp\left(-\frac{E_a}{RT}\right) N^2. \tag{XI-3}$$

From (XI-3) we obtain the time dependence of the concentration of defects:

$$N = \frac{N_{\text{in}}}{aN_{\text{in}} \exp\left(-\frac{E_a}{RT}\right)\tau + 1}. \qquad \text{(XI-4)}$$

Substituting (XI-4) into (XI-2), we obtain

$$\frac{dv}{v} = -\frac{bN_{\text{in}} \cdot \exp\left(-\frac{E_b}{RT}\right)}{aN_{\text{in}} \cdot \exp\left(-\frac{E_a}{RT}\right)\tau + 1} \cdot d\tau. \qquad \text{(XI-5)}$$

After integration, we have

$$\ln \frac{v}{v_{\text{in}}} = -\frac{b}{a} \cdot \exp\left(-\frac{\Delta E}{RT}\right) \ln\left[aN_{\text{in}} \cdot \exp\left(-\frac{E_a}{RT}\right)\tau + 1\right], \qquad \text{(XI-6)}$$

where $\Delta E_b - E_a$.

Equation (XI-6) described the variation of the change in volume of pores during sintering both with time and with temperature. At constant temperature it coincides with empirical equation (III-5).

Comparison of (XI-6) and (III-5), and alos (XI-5) and (III-4), leads to a more complete interpretationof constant q and m in Eq. (III-5). Since $q = (dv/d\tau)_{\text{in}} \cdot (1/v_{\text{in}})$ (the subscript indicates that the value refers to the beginning of isothermal sintering), in conformity with (XI-5), for $\tau = 0$ we have

$$q = bN_{\text{in}} \cdot \exp\left(-\frac{E_b}{RT}\right). \qquad \text{(XI-7)}$$

For convenience of comparison we write Eq. (XI-6) in the form

$$\frac{v}{v_{\text{in}}} = \left[aN_{\text{in}} \cdot \exp\left(-\frac{E_a}{RT}\right)\tau + 1\right]^{-\frac{1}{\frac{a}{b} \cdot \exp\frac{\Delta E}{RT}}}.$$

From comparison with Eq. (III-5)

$$\frac{v}{v_{\text{in}}} = (qm\tau + 1)^{-\frac{1}{m}}$$

it follows that

$$qm = aN_{\text{in}}\exp\left(-\frac{E_a}{RT}\right) \tag{XI-8}$$

and

$$m = \frac{a}{b}\cdot\exp\frac{\Delta E}{RT}. \tag{XI-9}$$

The latter relationship explains the experimentally established temperature dependence of the value of m. Since $E_b > E_a$ (see Chapter V), the sign of the exponent is positive and, consequently, m must decrease with increasing temperature.

In Chapter IV the phenomenological significance of m was defined as the intensity of the slowdown of densification with reduction of the concentration of defects. The analysis given above corroborates and defines m more precisely. In view of Eqs. (XI-2), (XI-3), and (XI-9), we can write

$$m = \frac{a}{b}\cdot\exp\frac{E_b - E_a}{RT} = \frac{aN\cdot\exp\left(-\frac{E_a}{RT}\right)}{bN\cdot\exp\left(-\frac{E_b}{RT}\right)} = \frac{\dfrac{dN}{d\tau\cdot N}}{\dfrac{dv}{d\tau\cdot v}}.$$

Consequently, m is equal to the ratio of the relative rates of reduction in the concentration of defects and reduction in the volume of pores. This means that with large m the reduction in volume of pores is accompanied by a sharp reduction in the concentration of defects. Since the rate of reduction in volume of pores is proportional to the concentration of defects, with large m there is an even sharper drop of the densification rate.

As was shown in Chapter IV, it is difficult to determine E_b in Eq. (XI-7) from the experimentally found values of q because of the changing value of product bN_{in} for the beginning of isothermal sintering at different temperatures. However, the value of E_b can be found by determining the rates of reduction in volume of pores before and after rapid change of the temperature following a more or less long period of isothermal sintering. If the change from one sintering temperature to another is rapid then the concentration of defects does not have time to change substantially. In this case the change in the rate depends mainly on the activation energy of bulk flow.

The rate of reduction in volume of pores at the end of the first period and at the beginning of the second period of isothermal sintering is determined in a dilatometric apparatus, by means of which the change in volume of pores is observed at both temperatures. The rates are determined graphically, by constructing tangents to the curves at the respective points. The rate of reduction in volume of pores is given per cubic centimeter of pores, since the volume of pores changes somewhat in the transition period from one temperature to the other.

The activation energy was determined by the standard formula

$$E_b = \frac{R}{0.434} \cdot \frac{\log \dot{v}_2 - \log \dot{v}_1}{\frac{1}{T_1} - \frac{1}{T_2}}, \qquad \text{(XI-10)}$$

where

$$\dot{v} = \frac{dv}{d\tau} \cdot \frac{1}{v}.$$

TABLE 25. Determination of the Flow Activation Energy E_b from Experimental Data Obtained with a Stepped Rise in Temperature*

Powder	Temperature, first sintering period, °C	$\dot{v}_1$ at end of first period, h^{-1}	Temperature, second sintering period, °C	$\dot{v}_2$ at beginning of second period, h^{-1}	E_b, cal/g-atom
Copper	850	0.0320	1000	0.881	62,700
Nickel	850	0.0258	1000	3.95	95,000
Silver	740	0.0185	850	1.54	73,000

*Isothermal sintering period 2 h, change from one temperature stage to the other in 4 min.

The results of determining E_b for copper, nickel, and silver, together with the original experimental data, are given in Table 25.

As would be expected, with the exclusion (or more precisely, the reduction) of the distorting effect of the change in the concentration of defects the activation energy of flow is higher than the values given in Chapter IV. It even exceeds the activation energy of self-diffusion – relatively little for copper and a great deal for silver and nickel.

Repetition of the experiment under the same conditions produced good agreement in the values of E_b (within the limits of ±2500 cal/g-atom). However, with a change of the temperature range in which $\mathring{v}$ is measured the value of E_b changes somewhat. With a lower temperature of the first sintering period the value of E_b usually decreases. This is due to the increase of the distorting effect of the change in the concentration of defects in changing from the temperature of the first to the temperature of the second period of isothermal sintering. If the temperature is relatively low a high concentration of defects is retained at the end of the first period. At the beginning of the following isothermal sintering the rate of reduction in volume of pores is determined not only by the change of temperature but also, to a considerable degree, by the reduction in the concentration of defects during the period of heating to the second isothermal sintering temperature. Therefore, in the range of relatively low temperatures the densification rate at the beginning of the second period decreases and the activation energy determined from these data will be below the true value with a constant concentration of defects. For example, when the measurements are made in the temperature range of 700-850°C the flow activation energy for copper is around 57,500 cal/g-atom, and 78,500 cal/g-atom for nickel. These values are considerably lower than those obtained at higher temperatures, where the distorting effect of the change in concentration of defects is smaller (Table 25).

The substantial role of the change in the concentration of defects is particularly manifest in the fact that the highest densification rate is frequently observed not in the beginning of resintering at a higher temperature but in the process of heating between the two periods of isothermal sintering. When the given temperature of the second isothermal sintering is attained the densification rate is already much lower. This is observed mainly in sintering of very active powders with a high concentration of defects.

Obviously, the distorting effect of changing concentrations of defects cannot be eliminated entirely. The distorting effect decreases with an increase of the first isothermal sintering temperature, with a relatively small difference in the first and second sintering temperatures, and of course with acceleration of the change from the first to the second isothermal sintering temperature. With this in mind, it is possible to select conditions in which the value approaches the true value for a hypothetical case of retaining a constant concentration of defects with a change in temperature. The indication that E_b is close to the true value is a small change in the value with further increase of the temperature or with a change in the heating rate between the periods of isothermal sintering.

The fact that the value of E_b calculated from experimental data with a stepped rise in temperature reflects the original temperature dependence of the flow rate at a constant concentration of defects is confirmed by the fact that these values turn out to be suitable for calculating the temperature dependence of densification and analyzing the change in the course of densification under different sintering conditions. In the calculations given below we use values of E_b found in experiments with a stepped rise in temperature.

The value of E_b varies with the method of obtaining the powder. The values of E_b determined under conditions ensuring the minimal dependence of the ratio of densification rates on the changing concentration of defects often but not always exceed the self-diffusion activation energy. Some powders (carbonyl nickel, for example) have an extremely low activation energy of flow. The value of E_b for carbonyl nickel is approximately one-half the self-diffusion activation energy of nickel.*

*The frequently observed high value of the flow activation energy of a defective crystal evidently points to the complex physical mechanism of this process. A phenomenologically elementary process (this term refers to a process with a constant activation energy obeying a given kinetic law) may be a complex phenomenon including several physically elementary processes. Under known conditions a combination of several physically elementary processes may have an apparently single activation ener-

The activation energy of another phenomenological elementary process – the elimination of defects – can be determined from experimental data characterizing the temperature dependence of coefficient m in Eq. (III-5). Using the value of m for several (at least two) temperatures, we find the value of ΔE (see below) by means of Eq. (XI-11). Then, subtracting ΔE from E_b, we find E_a:

$$E_a = E_b - \Delta E$$

(keeping in mind that the value of E_b was found earlier by means of an experiment with a stepped rise in temperature).

The activation energy of the elimination of defects varies with the method of obtaining the powder, i.e., with the formation conditions of crystals with defective lattices. For different powders of the same metal E_a may have substantially different values.

The variation of the reduction in volume of pores with temperature and time established by Eq. (XI-6) explains the results from many experiments in which the densification process was affected in one way or another by the temperature. Calculations with Eq. (XI-6) require knowledge of the value of its constant. The values of E_b and E_a are found by the method described above. The value of b reflects the value of that characteristic of the initial porosity which remains unchanged at different compacting pressures (within the limits of constant v_s/v_p at different d_p), and therefore the value of b, and also the value of a, is constant for a given powder. The absolute values of a and b are difficult to determine, since there is no possibility of determining the dimensions and value of N_{in}. However, this is unnecessary in the present case. The ratio a/b is found together with the value of ΔE from the values of m experimentally determined for a given powder at several temperatures. In conformity with Eq. (XI-9), with known

---

gy. A physical analysis of the flow process in a defective crystal probably requires additional work on the problem of the defective condition of the crystal lattice, which is outside the limits of this study. A phenomenological analysis avoiding the question of the physical nature of the phenomenon is based on the experimentally established possibility of describing the temperature dependence of the elementary processes by means of activation energies that are constant for a given powder. Additional considerations on the complex nature of phenomenologically elementary processes can be found in Chapter XIII.

values of m_1 and m_2 for temperatures T_1 and T_2 we find

$$\Delta E = \frac{R}{0.434} \frac{\log m_1 - \log m_2}{\frac{1}{T_1} - \frac{1}{T_2}} \quad \text{(XI-11)}$$

and

$$\frac{a}{b} = m_{1(2)} \exp\left(-\frac{\Delta E}{RT_{1(2)}}\right). \quad \text{(XI-12)}$$

More precise values of ΔE and a/b can be found by plotting $\log m$ vs. $1/T$. Since the relationship is linear, a straight line is drawn along the experimental points. On the straight line we select two points with coordinates $\log m_1$, $1/T_1$ and $\log m_2$, $1/T_2$. These values are used for calculating ΔE and a/b by Eqs. (XI-11) and (XI-12).

Thus, by a relatively simple experimental method (one sintering with a stepped rise in temperature, several sinterings for 2 h, and plotting v_s/v_p vs. τ for different temperatures) one can find the constants E_b, E_a, and a/b, knowledge of which makes it possible to calculate the course of densification at different temperatures. Some complication is introduced by the fact that aN_{in} enters into Eq. (XI-6), which is not a constant of a powder. As we have said many times, the concentration of defects at the beginning of isothermal sintering depends on the heating period and also the temperature range of heating to the beginning of isothermal sintering. Therefore, the value of aN_{in} must conform with the given conditions at the beginning of isothermal sintering. To calculate aN_{in} it is necessary to know constants q and m in Eq. (III-5). Let us assume that they are found from the three selected points on the initial section of the curve of v_s/v_p vs. τ, as was described in Chapter III. From Eq. (XI-8) it follows that

$$aN_{in} = qm \cdot \exp\frac{E_a}{RT}. \quad \text{(XI-13)}$$

The dimensions of aN_{in} are relative to the rate, i.e., h^{-1}.*

*The dimensions of aN_{in} come from Eq. (XI-3):

$$aN = \frac{dN}{d\tau} \cdot \frac{1}{N} \exp\frac{E_a}{RT} \; h^{-1}.$$

With determined values of E_b, E_a, and a/b the value of aN_{in} can be calculated even if only one value of v_1/v_{in} for time τ_1 is known. By means of $\Delta E = E_b - E_a$ and a/b we find the value of m for a given temperature by Eq. (XI-9). It follows from (III-5) that $qm = [(v_{in}/v_1)^m - 1] \cdot (1/\tau_i)$, and together with (XI-13) we have

$$aN_{in} = \frac{\left(\frac{v_{in}}{v_1}\right)^m - 1}{\tau_1} \exp \frac{E_a}{RT}. \qquad \text{(XI-14)}$$

Thus, the value of aN_{in}, depending on the conditions of heating to the beginning of isothermal sintering, can be found when the coordinates of the point corresponding to the beginning of isothermal sintering (v_{in} at $\tau = 0$) are known and any of the points on the isothermal curve on plots of v_s/v_p vs. τ.

Thus, of the four constants of Eq. (XI-6), three are constants of the powder, while the fourth, aN_{in}, is determined both by the properties of the powder and the heating conditions up to the beginning of isothermal sintering. When all four constants are known then the course of densification can be calculated for single isothermal sintering, and also for more complex cases. Thus, one can calculate the reduction in volume of pores attained after double sintering and compare it with the reduction in volume of pores after single sintering conducted under indentical conditions with subsequent high-temperature sintering.

In solving similar problems it must be understood that the concentration of defects, and along with it the value of aN_{in} for the following section of the isothermal curve (in repeated sintering), will decrease substantially because of the reduction in the concentration of defects in the first sintering period. The change in the value of aN due to the disappearance of some of the defects can be determined by Eq. (XI-4). Multiplying both sides of this equation by a, we obtain an equation making it possible to calculate the relative concentration of defects, expressed by aN, after sintering time τ:

$$aN = \frac{aN_{in}}{aN_{in} \exp\left(-\frac{E_a}{RT}\right)\tau + 1}. \qquad \text{(XI-15)}$$

If the length of the first period of multistage sintering is equal to τ, then the value of aN for the end of the first period can be calculated by (XI-15). This value is taken as the value of aN_{in} (initial concentration of defects) for the second period of isothermal sintering. The change in the concentration of defects in passing from the temperature of the first to the temperature of the second stage of isothermal sintering can be neglected; this is permissible if the transition is completed fairly rapidly. The values of aN_{in} for many subsequent stages of isothermal sintering can be calculated in a similar way.

Let us now make the calculations for the specific case of sintering of a powder with known values of E_b, E_a, and a/b. These values for copper powder obtained by reduction of oxides in hydrogen at 400°C are as follows: E_b = 62,700 cal/g-atom; E_a = 48,400 cal/g-atom; a/b = 1.095×10^{-2}. It was also established that v/v_{in} = 0.863 after sintering 1 h at 700°C. Using Eq. (XI-13), we find that $(aN_{in})_1 = 2.82 \times 10^{11}$ h^{-1} for these sintering conditions. The value of $(aN_{in})_2$ for the following sintering period, calculated by (XI-15), is equal to 1.12×10^{10} h^{-1}. Using (XI-6), we find v/v_{in} for the end of the second isothermal sintering period at 900°C for 1 h. It is equal to 0.542. The total densification after the two periods of isothermal sintering at 700 and 900°C will equal

$$\left(\frac{v}{v_{in}}\right)_{1+2} = \left(\frac{v}{v_{in}}\right)_1 \cdot \left(\frac{v}{v_{in}}\right)_2 = 0.863 \cdot 0.542 = 0.468.$$

Using the same constants of the powders, let us calculate v/v_{in} for single sintering at 900°C for 1 h. The value of aN_{in} is arbitrarily taken as equal to the value found for 700°C. Then, substituting all the values of the constants and also the time (τ = 1) and temperature (T = 1173°K) into Eq. (XI-6), we obtain v/v_{in} = 0.289.

Thus, the calculations show that the reduction in volume of pores after single sintering at 900°C must exceed the total reduction in volume of pores for sintering twice (first at 700 and then at 900°C), which agrees with the experimental data (see Chapter V).

Quantitative agreement with the experimental data is difficult to obtain, however, since it is practically impossible to heat the samples to 900°C in such a manner that the concentration of defects and also the value of v_{in} at the beginning of isothermal sintering will coincide with the values for heating to 700°C. A more rigorous

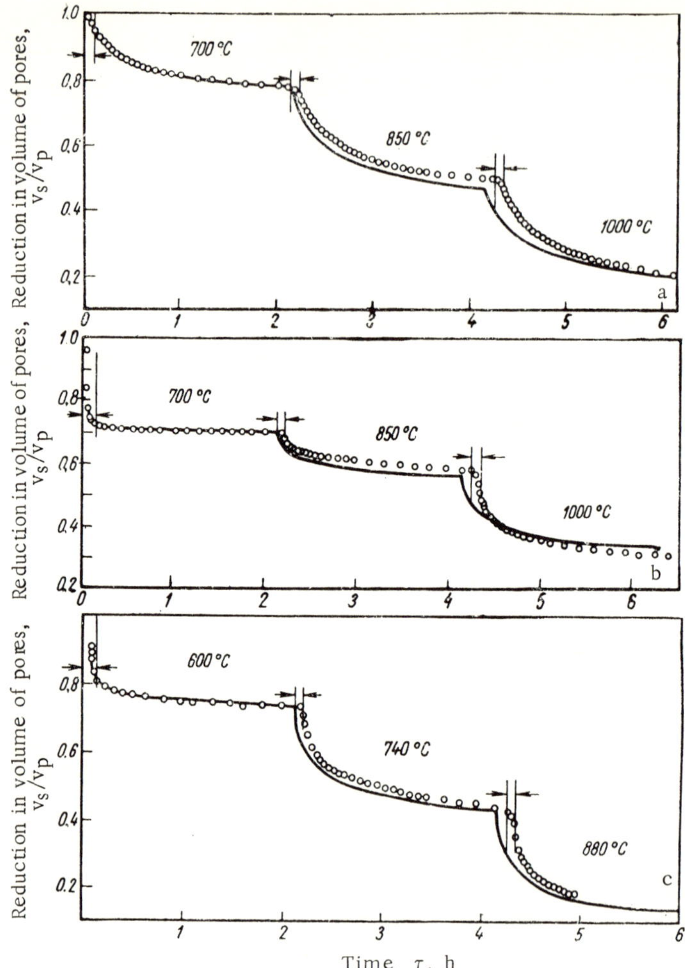

Fig. 30. Calculated (solid lines) and experimental data (points) on v_s/v_p vs. τ in three periods of isothermal sintering of copper (a), nickel obtained by reduction of oxides in hydrogen (b), and silver (c) with temperatures increasing one after the other (experiment with stepped rise in temperature). The arrows indicate the periods where the temperature rises.

test of Eq. (XI-6) is to calculate the change in the value of v_s/v_p for several periods of isothermal sintering with increasing temperatures and compare the theoretical curves with the experimental data. In the calculation we use the constant of the powder E_b, E_a, and a/b, and also the kinetic characteristic of the relative concentration of defects at the beginning of the first sintering period $(aN_{in})_1$ determined by calculation with the same value of v_1/v_{in} for some time τ_1 in the first period of isothermal sintering. The variation of v_s/v_p in the following periods of isothermal sintering at other temperatures will be found by calculation.

Control experiments were made in the dilatometric apparatus described above (Chapter III) with rapid heating to sintering temperature and rapid change from one isothermal sintering temperature to another. Copper and nickel powders were sintered at temperatures of 700, 850, and 1000°C, with holding at each temperature for 2 h. Silver powder was sintered at 600, 740, and 880°C, for 2 h at each temperature. The change from one temperature stage to another was completed in approximately 4 min.

Figure 30 shows the variation of v_s/v_p with τ from the experimental data and the data calculated by Eq. (XI-6) for all the isothermal sintering periods. The value of $(aN_{in})_2$ for the following stage was found from the value of $(aN_{in})_1$ of the preceding stage by using Eq. (XI-15). The theoretical curves for the second and third sintering periods should closely approximate the experimental values for all three experiments. Although complete agreement was not obtained (naturally, since the changes occurring in the heating period between isothermal sintering periods were disregarded), the shape of the curves, reflecting the deceleration of the process, is quite similar. In particular, there is little difference in the shape of the curves or in the densification (reduction in volume of pores) in the second and third periods of sintering of copper, and the noticeable differences in the curves of the second and third stages for nickel (the change in v_s/v_p in the third stage was double that in the first) are predicted exactly by the calculations. Thus, the calculations by Eq. (XI-6) correctly reflect both the time and temperature dependence of densification.

For the powders in these experiments the kinetic constants used in the calculations were as follows:

	E_a, cal/g-atom	E_b, cal/g-atom	a/b	aN_{in}, h^{-1}
Copper	48,400	62,700	1.095×10^{-2}	8.98×10^{11}
Nickel	72,000	95,000	8.06×10^{-4}	3.91×10^{17}
Silver	54,000	73,000	1.15×10^{-3}	2.44×10^{13}

Experiments with a stepped increase in temperature graphically demonstrate the single mechanism of the process for the beginning of sintering with low densification (in the first stage) and at higher temperature stages with a substantial reduction of the porosity toward the end of sintering. The form of the curves gives no reason to assume a change in the mechanism of densification in the process of sintering or to assume the continuous action of associated processes preventing densification. In particular, breaking and shifting of contacts, which were regarded as one of the continuously acting factors determining the kinetics of densification in [2, 131, 132], evidently occur only during heating before the beginning of isothermal sintering. This results from the fact that the kinetics of densification at the third temperature stage, with considerably strengthened contacts, is the same as in the first stage, at the beginning of which the breaking of contacts is still probable to some extent. Evidently the elastic-plastic aftereffect, breaking the contacts, develops mainly in the process of raising the temperature. After the beginning of isothermal sintering its effect on the course of densification is almost unnoticeable.

A rapid drop of the densification rate in each temperature stage of isothermal sintering is characteristic for sintering of crystalline bodies. In this case a basic characteristic of the temperature dependence of the process is apparent – an increase of temperature accelerates not only the flow of the substance but also the elimination of defects. The rate of the decrease in the densification rate and the extent of the change in v_s/v_p within the limits of each temperature stage depend on the relationship of the activation energies of both processes. Of course, these phenomena cannot be observed in sintering of amorphous bodies (Fig. 31). Here, the flow accelerates during the transition to a higher temperature

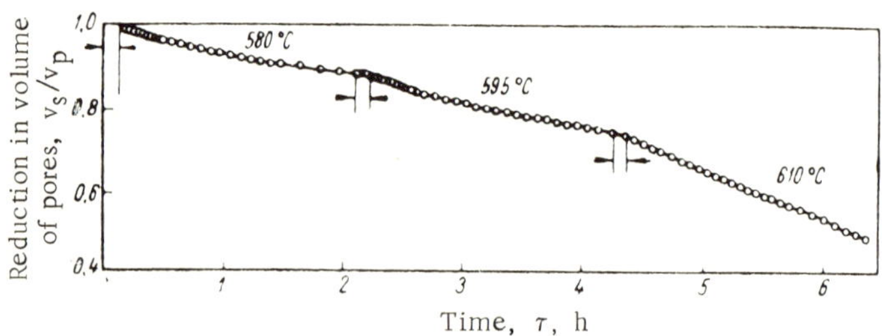

Fig. 31. Variation of v_s/v_p with τ during three isothermal sintering periods (each at a higher temperature) for glass powder.

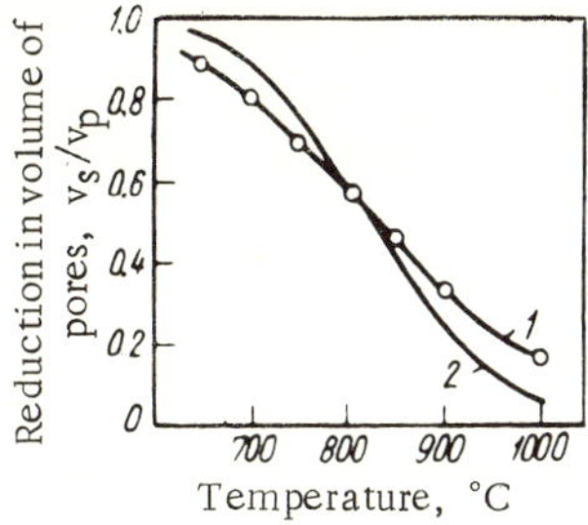

Fig. 32. Temperature dependence of v_s/v_p for sintering of copper compact with holding at constant temperature 1 h. 1) Experimental curve; 2) curve calculated by Eq. (XI-6).

stage, and within the limits of each isothermal sintering period the decrease of the rate is very small, which is explained by the absence of a change in flow with time that is characteristic of crystalline bodies. Comparison of the forms of the curves for sintering of metallic and glass powders in experiments with stepped heating gives a clear idea of the differences in the kinetics of densification for crystalline and amorphous bodies (Figs. 30 and 31).

The use of Eq. (XI-6) to describe the variation of the reduction in volume of pores with temperature at a constant sintering time is complicated by the variable value of aN_{in}, decreasing with increasing isothermal sintering temperatures, since the concentration of defects in the initial heating period decreases at a greater rate with increasing sintering temperatures. Another difficulty results from the fact that the volume of pores at the beginning of isothermal densification also depends greatly on the sintering temperature (let us remember that densification during heating to the isothermal sintering temperature usually constitutes a substantial part of the total densification even with prolonged isothermal sintering).

If we disregard the fact that these values are not constant and calculate v_s/v_p as a function of temperature, taking $v_{in} = 1$ and the value of aN_{in} found for some temperature in the middle of the temperature range we are interested in, we obtain a curve intersecting the experimental curve at a point corresponding to the sintering temperature for which the value of aN_{in} was found. Figure 32 shows the experimentally determined relationship of $v_s/v_p = \varphi(T)$ for copper powder sintered 1 h and followed by a rise in temperature after 10 min, and also the curve plotted from data calculated with use of the constants found in experiments with steeped heating (Fig. 30a). The value of aN_{in} for 800°C, determined by

Eq. (XI-14), amounted to 7.73×10^{11} h^{-1}. The "slewing" of the experimental curve with respect to the curve resulting from calculations by Eq. (XI-6) is natural and unavoidable, since it reflects the changes occurring during heating to the isothermal sintering temperature. At temperatures below 800°C the concentration of defects at the beginning of isothermal sintering is higher than at 800°C. Therefore densification occurs at a high rate and the densification attained is higher than indicated by the calculation with a low (for these temperatures) value of $aN_{in(800)}$. The reverse situation occurs at temperatures above 800°C: Here the reduction of the concentration of defects in the initial heating period is higher than during heating to 800°C, and the actual value of aN_{in} is smaller than $aN_{in(800)}$, which reduces densification (reduction in volume of pores) as compared with the calculations.

Obviously, the distorting effect of nonidentical changes in the concentration of defects during the heating period is unavoidable and uncorrectable with this method of calculation, although the distortion decreases with increasing initial heating rates. Nevertheless, the curves of v_s/v_p vs. T plotted from the constants of the powder E_b, E_a, and a/b and from aN_{in} for some medium temperature correctly reflect not only the general character of the change of v_s/v_p with temperature but also the effect of the properties of the powder or sintering conditions on the form of these relationships. Let us take as an example the effect of the temperature at which the metal powder is obtained and subsequently calcined (before compacting) on the character of relationship $v_s/v_p = \varphi(T)$. This effect was investigated on cobalt powder obtained at different temperatures by reduction of oxides in hydrogen. Part of the powder was subjected to additional calcining at different temperatures. The compacts were sintered at different temperatures for 0.5 h. Heating to the given temperature took about 10 min. The variation v_s/v_p with T plotted from the experimental data is shown in Fig. 33. All the curves have a similar shape. With an increase of reducing or calcining temperatures the curves shift toward higher temperatures, the slope remaining almost unchanged in the middle section.

Now, departing from the experimental data obtained, let us consider what conclusions on the change in the character of relationship $v_s/v_p = \varphi(T)$ can be drawn by taking into account the effect of

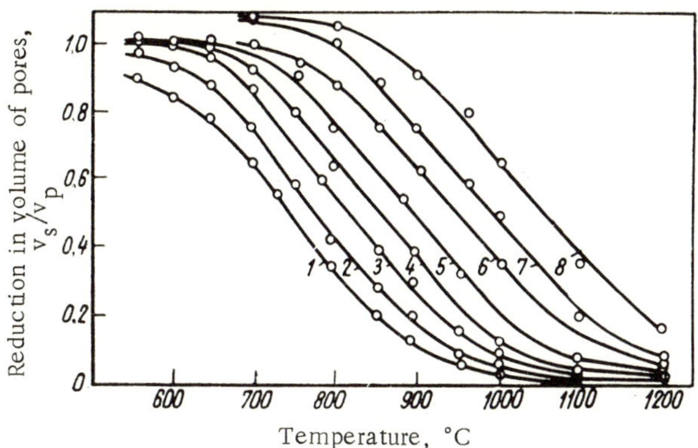

Fig. 33. Temperature dependence of v_s/v_p in sintering cobalt powder for 0.5 h. 1) Powder obtained by reduction of oxides in hydrogen at 525°C; 2) at 580°C; 3) at 625°C; 4) at 675°C; 5) at 750°C; 6) powder obtained by reduction at 750°C and calcined at same temperature; 7) same, calcined at 800°C; 8) same, calcined at 900°C.

calcining the powder on the value of the constant in Eq. (XI-6). Let us remember that the derivation of this equation is based on the assumption that the process of elimination of defects is independent of the size and shape of the particles or pores. The elimination of defects in the compact and in the freely poured powder must obey the same laws. Hence it follows that a change in the concentration of defects during calcining of the powder can be described by Eq. (XI-4). On the basis of the same analogy, it must be assumed that the activation energy of the elimination of defects will vary only with the primary origin (method of obtaining the powder) and not with the calcining temperature, exactly in the same way as it is independent of sintering temperature. These considerations make it possible to use the following form of Eq. (XI-15) to calculate the change in the concentration of defects during calcining of the powders:

$$aN_{in} = \frac{aN_0}{aN_0 \cdot \exp\left(-\frac{E_a}{RT}\right)\tau + 1}, \qquad \text{(XI-16)}$$

where T and τ are the temperature and time of calcining the powder, aN_{in} is the relative concentration of defects after calcining (or in the beginning of subsequent sintering), and aN_0 is the relative concentration of defects in the original powder before calcining (it is assumed that heating to calcining temperature occurs instantaneously).

By means of Eq. (XI-16) we find the value of aN_{in} after calcining of the powder at different temperatures.

The other constants in Eq. (XI-6) do not change or change little with calcining. The fact that E_a and E_b are constant requires no explanation, since it is the main premise of the interpretation proposed for the kinetics of densification. The value of b in ratio a/b changes negligibly at moderately high calcining temperatures. Let us recall that b is a function of the initial pore geometry, which is determined mainly by the size of the particles and is almost independent of the surface relief or the shape of the pores (more detailed considerations will be given below). If calcining of the powder does not lead to growth of the particles but results mainly in smoothing of the surface and simplification of the shape, then b remains constant. This is confirmed by experiment: As already noted in Chapter IV, calcining of the powder produces little or no change in the value of m in Eq. (III-5), which, in conformity with Eq. (XI-9), is proportional to a/b at a given temperature.

With a slight increase of the calcining temperature as compared with the temperature at which the powder was obtained there is almost no growth of the particles and the changes amount to smoothing of the surface relief, growth of grains within particles, and a change in the concentration of defects. When it is necessary to increase the calcining temperature substantially, the following method is usually used to avoid sintering of the particles: Calcining is conducted at successively higher temperatures, the powder being cooled and pulverized after each calcining in order to break the bonds between particles. This method can be used to obtain powders that change little in particle size even after sintering at relatively high temperatures. This method was used in the experiment described above with cobalt powder (Fig. 33).

If only the value of aN_{in} changes during calcining, then, in conformity with the variation of v_s/v_p with aN_{in} established by Eq. (XI-6), the change in aN_{in} at all sintering temperatures will induce a change in v_s/v_p in the same direction: v_s/v_p increases with decreasing aN_{in}. It follows from this that curves of v_s/v_p =

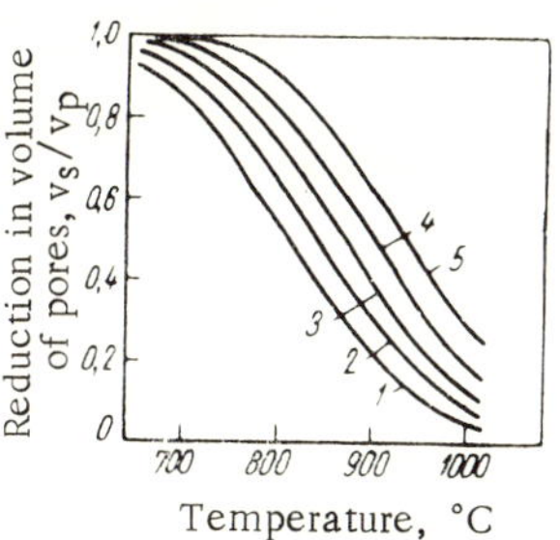

Fig. 34. Temperature dependence v_s/v_p calculated for isothermal sintering 1 h of copper powders calcined at different temperatures. 1) Original powder; 2) powder calcined at 650°C; 3) at 700°C; 4) at 750°C; 5) at 800°C.

$\varphi(T)$ for powders with different preliminary calcining temperatures will not intersect, since all points on the same curve will shift in the same direction with respect to each other. This can be shown more graphically by calculating a set of curves of $v_s/v_p = \varphi(T)$ for several powders calcined at different temperatures. Since there were no data on the constants of the powders used in the experiment described, the calculations given below were made for copper powder whose constants were found in the experiment with stepped heating (Fig. 30a). From the value of E_a and $aN_0 = 9 \times 10^{11}\ h^{-1}$ we found the value of aN_{in} by Eq. (XI-16) for powders calcined for 1 h at 650, 700, 750, and 800°C. Then, using $E_a = 48{,}400$ cal/g-atom, $E_b = 62{,}500$ cal/g-atom, and $a/b = 1.064 \times 10^{-2}$ from the same experiment, we plotted the variation of v_s/v_p with T, using Eq. (XI-6), for 1 h of sintering. The set of curves is shown in Fig. 34. Comparison with Fig. 33 shows that the calculations correctly

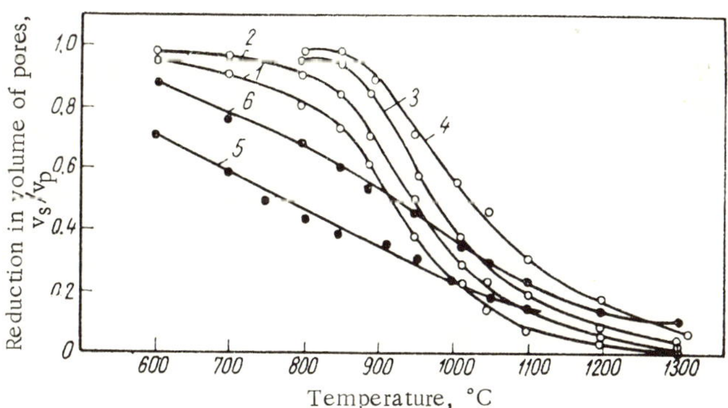

Fig. 35. Temperature dependence of v_s/v_p for nickel powders obtained by different methods and held at sintering temperature 0.5 h. 1) Powder obtained by reduction of oxides in hydrogen at 650°C; 2) same, with multiple calcining, the last time at 700°C; 3) same, the last time at 800°C; 4) same, the last time at 900°C; 5) carbonyl nickel; 6) same, calcined at 500°C.

reflect the change in the position of the curves at high calcining temperatures: The curves hardly change in shape, but are shifted to higher temperatures.

The curves in Fig. 33 for both the calcined powders and also the powders obtained at different temperatures shift in relation to each other with no notable change in slope. Evidently high-temperature reduction can be regarded, with well-known approximations, as a combination of low-temperature reduction with subsequent calcining. This assumption is undoubtedly justified with a slow increase of temperature in the process of reduction. The same slope of the curves in the plot of v_s/v_p vs. T points to equal values of E_a and E_b for all powders.

The arrangement of the curves is different in plots of v_s/v_p vs. T for powders obtained by different methods if they have sharply differing activation energies of elementary processes. Figure 35 shows $v_s/v_p = \varphi(T)$ for two groups of nickel powders obtained by reduction of oxides in hydrogen and by dissociation of carbonyl. Part of the powder was subjected to additional calcining in hydrogen. The curves for powders obtained by the same method and differing only in their calcining temperatures have almost the same slope, while the curves for powders obtained from carbonyl and from oxides have different slopes, and the curves intersect. It is not difficult to show that the difference in slope is due directly to the effect of the value of the activation energy of the elementary processes on the shape of the curves from Eq. (XI-6). It follows from

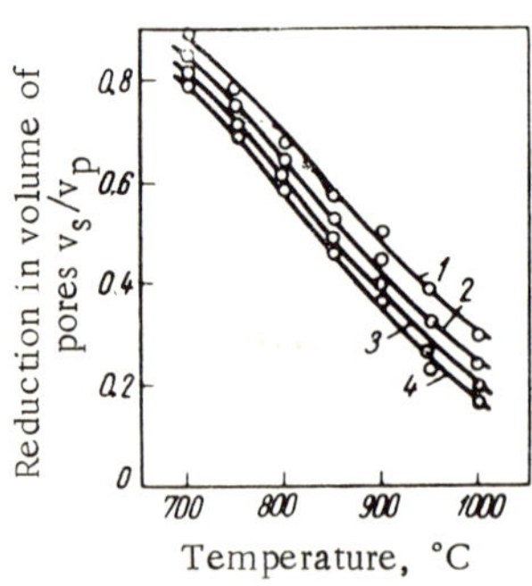

Fig. 36. Temperature dependence of v_s/v_p calculated for isothermal sintering for 1 h from the constants of nickel powders obtained by different methods. 1) Powder obtained by dissociation of carbonyl (E_a = 18,100 cal/g-atom; E_b = 28,400 cal/g-atom; $a/b = 3.77 \times 10^{-2}$; $aN_{in} = 7.5 \times 10^5$ h^{-1}); 2) powder obtained by reduction of oxides in hydrogen (E_a = 72,000 cal/g-atom; E_b = 95,000 cal/g-atom; $a/b = 8.06 \times 10^{-4}$; $aN_{in} = 4.1 \times 10^{18}$ h^{-1}).

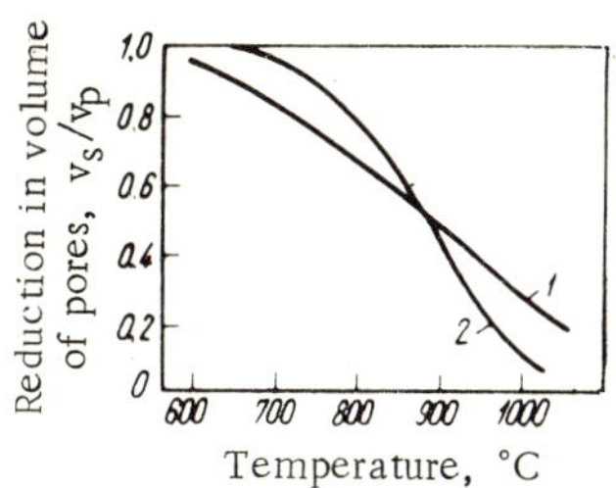

Fig. 37. Temperature dependence of v_s/v_p for copper powder with different sintering times. 1) 0.5 h; 2) 1 h; 3) 2 h; 4) 3 h.

analysis of this equation that the slope of $v_s/v_p = \varphi(T)$ increases if the value of ΔE or E_a increases. Therefore the curves for powders obtained by different methods often have a different slope and may intersect. Figure 36 shows curves of $v_s/v_p = \varphi(T)$ calculated from the previously found constants for two nickel powders – reduced from oxides and obtained by dissociation of carbonyl (the latter additionally calcined in hydrogen). The calculation produced intersecting curves, the location of which was similar to that of curves plotted from the experimental data (Fig. 35).

Thus, Eq. (XI-6), not providing complete agreement with the experimental curve for reasons given above, explains the difference in the shape of the curves for powders obtained by different methods. With known values of constants E_a, E_b, and a/b, Eq. (XI-6) makes it possible to determine semiquantitatively the effect of preliminary calcining on the position of curves of $v_s/v_p = \varphi(T)$.

In conformity with (XI-6), a change of sintering time should change the position of curves of $v_s/v_p = \varphi(T)$, in the same manner as for preliminary calcining of the powders. This results from the structure of the equation: The values aN_{in} and τ are cofactors in the product $aN_{in} \cdot \exp(-E_a/RT)\,\tau$, and therefore a change in either value has an identical effect on the shape of the curves. This is fully confirmed by experiment. Figure 37 shows the change in the position of curves of $v_s/v_p = \varphi(T)$ with different sintering times for copper powder. With an increase of sintering time the curves shift to the left, the slope remaining almost unchanged. With calcining of powders, the value of aN_{in} decreasing, the curves shift to the right, again with hardly any change in slope (Fig. 33). This is an indirect confirmation of the assumption that with phenomenological description of the process one can at least derive from the constant value of the activation energy of elimination of defects E_a

that is independent of the concentration of defects and changes with calcining of the powders or sintering of compacts. Analysis of Eq. (XI-6) shows that with a change of E_a after calcining the slope of the curves would inevitably change (as is confirmed by comparison of curves v_s/v_p vs. T for powders differing in origin, Fig. 35).

Thus, Eq. (XI-6) can give a correct estimate of the character of the effect of changing sintering conditions on the form of relationship $v_s/v_p = f(\tau)$ with T = const and $v_s/v_p = \varphi(T)$ with τ = const. The time dependence is quantitative and the temperature dependence semiquantitative.

Complete agreement of the calculations with the experiment is prevented by the fact that aN_{in} is not constant under actual sintering conditions (aN_{in} changes because of gradual heating instead of the instantaneous rise in temperature at the beginning of sintering assumed in the calculations). Generally speaking, in view of the temperature and time dependence of elementary processes one can pose the problem of calculating the course of densification with continuously changing temperature. The problem can be regarded as real and can be solved by means of integrating the respective differential equations. Such calculation would make it possible to determine the change of aN and v_s/v_p during heating to a given temperature at a constant heating rate with an instantaneous change to constant temperature. Although this ideal change in temperature at the beginning of sintering is unreal, it is considerably closer to the usual course of heating than the case of instantaneous increase in temperature. However, exact calculation of densification and the change in aN in the period of the rise in temperature is fairly complicated. Only approximate calculations of the change in the value of aN and v_s/v_p in the period of heating to isothermal sintering temperature are of practical value (the means and the possibility of approximate calculations will be considered in the following chapter).

Analysis of the experimental data characterizing densification with a rise in temperature showed that kinetic equation (XI-5) is suitable also for describing densification with a varying temperature. This follows from the fact that the experiments described in Chapter V revealed the proportionality of the densification rate and the rate of increase in temperature: The rate of reduction in volume of pores increased in direct proportion to the increase in

TABLE 26. Variation of Rate of Reduction in Volume of Pores with Rate of Increase in Temperature

Powder	Temperature range, °C	Rate of increase in temperature, α, deg/min	Average rate of reduction in volume of pores,* $\dot{v}$, cm³/min	$\dot{v}/\alpha \times 10^3$, cm³/deg
Copper	800—1000	13	0.023	1.77
		34	0.056	1.65
		71	0.125	1.76
		266	0.256	0.97
Nickel	600—1000	14	0.0136	0.97
		36	0.0356	0.99
		57	0.0591	1.03
		100	0.1044	1.04

*With v_p = 1 cm³.

heating rate. Additional data characterizing this relationship are given in Table 26.

The ratio of the rates of reduction in volume of pores and the increase of temperature for each powder prove to be almost constant (the only exception being sintering of copper powder with the highest heating rate). The small variations of the ratio are probably due to a deviation of the rise in temperature with time from a linear relationship.

The constant value of the ratio is a direct consequence of the variation of the reduction in volume of pores with the rates of elementary processes, the temperature dependences of which are determined by Eqs. (XI-2) and (XI-3). It follows from (XI-2) that the rate of reduction in volume of pores, per unit volume of pores, is proportional to the concentration of defects at a given temperature:

$$\frac{dv}{d\tau} \cdot \frac{1}{v} \sim N.$$

Therefore the problem of determining the rate of reduction in volume of pores after heating amounts to determining the concentration of defects at the time the given temperature is attained. The change in the concentration of defects with time is described by Eq. (XI-4). After several transformations this equation can also be used to describe the change in the concentration of defects with sintering at a constant rate of the increase in temperature. Replacing τ in Eq. (XI-3) with $(T-T_0)/\alpha$ (where T_0 and T are the temperatures at the beginning and end of heating, α is the average rate of the increase in temperature) and replacing $d\tau$ with dT/α, we obtain after integrating in limits of T_0 and T

$$N = \frac{N_0}{\frac{aN_0}{\alpha}\int\limits_{T_0}^{T} \exp\left(-\frac{E_a}{RT}\right) dT + 1} \quad \text{(XI-17)}$$

As a rule, for active powders with a high concentration of defects

$$aN_0 \int\limits_{T_0}^{T} \exp\left(-\frac{E_a}{RT}\right) dT \gg 1.$$

When the value of α is not too high the units in the denominator can be neglected. Then Eq. (XI-17) is simplified and takes the form

$$aN = \frac{\alpha}{\int\limits_{T_0}^{T} \exp\left(-\frac{E_a}{RT}\right) dT}. \quad \text{(XI-18)}$$

The value of the denominator in this expression for a given powder is determined only by the initial and final temperature and does not depend on the heating time. Consequently, after heating within limits of T_0 to T the concentration of defects, and along with it the densification rate (or reduction in volume of pores), should be proportional to α, the rate of increase in temperature, which is fully confirmed by experiment.

A constant value of $\dot{v}/\alpha$ is characteristic of sintering crystalline bodies. It is not characteristic of sintering amorphous bodies, the densification rate of which with a given geometry of pores is determined only by the temperature at the time the rate is measured and not by the preceding heating time, and consequently it is independent of the rate of increase in temperature.

The proportionality of the densification rate and the rate of increase in temperature does not hold true for sintering low-activity crystalline powders (with small aN_0) or for sintering any powders with very high heating rates. In these cases it is impossible to neglect the units in Eq. (XI-17). The small value of $\dot{v}/\alpha$ for copper powder with α = 266 deg/min (see Table 26) is explained by this particular circumstance.

To complete the phenomenological analysis of the basic laws of densification we must determine the value of the constant of the powder a/b more precisely. Let us recall the origin of constant b: It is the product of the pre-exponential factor ϑ in Eq. (XI-1) and coefficient β in Eq. (XI-4), i.e., $b = \vartheta\beta$. The first of these factors depends on the nature of the defects and the substance of the body being sintered, and the second (coefficient β) determines the dependence of the rate of reduction in volume of pores on the volume of pores with a given structure (form and size) of the pores with constant flow. It follows that constant β is equal in value in the description of the reduction in volume of pores in amorphous and crystalline bodies: It characterizes the effect of the initial geometry of the pores on the subsequent course of densification, which for both amorphous and crystalline bodies is determined by Eq. (IX-4) with the sole difference that γ is constant for amorphous bodies and variable for crystalline bodies.

The value of β in (IX-4) can be determined more exactly by means of comparing this equation with the corresponding equation obtained from the theory of viscous flow. The theoretical concepts of Frenkel' hold true unconditionally for any isotropic viscous substance with constant flow, and it follows that for a body with unconnected cylindrical channel-like pores the variation of the rate of reduction in volume of pores with the volume of pores must be described as follows:

$$\frac{dv}{d\tau} = -\gamma \cdot 2\pi^{\frac{1}{2}} \sigma n l^{\frac{1}{2}} v^{\frac{1}{2}}, \qquad \text{(XI-19)}$$

where σ is surface tension, n is the number of pores (channels) per unit mass of the sintered body, l is the average length of channels, and v is the current value of the volume of pores.

Let us compare this equation with (IX-4):

$$\frac{dv}{d\tau} = -\gamma\beta v.$$

The identical structure of the two formulas is evident. The difference in the exponents of the volume is explained by the difference in the geometric characteristic of the pores – in one case (hypothetical) cylindrical pores with identical curvature of the surface in all sections, and in the other case (real body) pores in the form of a network of intersecting channels with different curvatures in different sections, including sections with both positive and negative curvatures (let us recall that the direct proportionality of $dv/d\tau \sim v$ was confirmed both by the observed kinetics of densification of amorphous bodies and the constant value of v_s/v_p in sintering bodies with different densities of the metal powders – see Chapter IX).

The product $2\pi^{1/2}\sigma n l^{1/2}$ in Eq. (XI-19) matches the value of β in Eq. (IX-4). This is also the characteristic of the geometry of pores, only very simplified with respect to the hypothetical case of cylindrical pores. In first approximation the product is constant, since it includes only that characteristic of the geometry of pores which is almost unchanged with reduction in volume of pores (the length of channels changes only slightly in the process of sintering). The value of the surface tension enters into the product. It is obvious that β is also directly proportional to the surface tension of the substance being sintered. The values of the third factor – flow, γ – coincide in both equations.

It follows from the comparison that for a body with cylindrical intersecting pores the coefficient equivalent to β is proportional to n (the number of pores per unit mass of the body). The question arises as to how β will change with complication of the shape of the pores in real bodies and, in particular, whether β could not be proportional to the specific surface of pores at a given volume of pores. From the constant reduction in volume of pores at different d_p it follows that the latter proposition cannot be correct. With increasing density of the body the ratio of surface to volume increases (this, in particular, is easily determined by measuring the adsorption capacity and the volume of the compact with successively increasing compacting pressures). Despite the large value of the surface per unit volume of pores, proportionality $dv/d\tau \sim v$ holds true with an increase of density. Consequently, β varies primarily with that geometric characteristic of the pores which does not change with increasing compacting pressure and increasing density of the compact. This characteristic is the value n, the number of pores per unit mass of the body. With regard to the actual structure of porous bodies, in many cases the value of n can be considered proportional to the number of particles per unit mass of the powder.

Thus, constant β is determined above all by the size of the particles or dispersity of the powder, but not its specific surface. Evidently, the fine structure of the surface of the pores is not reflected directly by the rate of flow in the entire volume of the body being sintered. Small unevennesses in the surface are eliminated by local processes not directly affecting the reduction in volume of pores. This follows from the constant relative reduction in volume of pores and at the same time gives some explanation of this relationship.

If, however, a connection between the fine structure of the surface and the flow in the bulk of the particles of the sintered body exists for all that (this equation is discussed in Chapter XIII) then the connection must indicate an effect on the course of densification regardless of the relative value of the pore surface or the relationship between the surface and volume of pores (in the contrary case v_s/v_p would not be constant at different d_p).

Thus, constant β varies mainly with the dispersity of the powder. Constant $b = \vartheta\beta$ varies with the pre-exponential factor ϑ in Eq. (XI-1), relating the flow rate with the concentration of defects. Since the value of ϑ varies with the type of defects, it will be different for powders obtained by different methods. For powders containing a given type of defect the value of ϑ is the same, and constant b varies only with the dispersity of the powder. This makes it possible at least qualitatively to predict the change in the shape and position of curves $v_s/v_p = f(\tau)$ with changes in the dispersity of the powder. Equations (XI-7) and (XI-9) define the connection between constants q and m in Eq. (III-5) and the value b. With increasing b, constant q increases proportionally with b, while m decreases with increasing b. Sintering experiments with powders of brittle substances that are easily ground (carbides, for example) show that after refining of the powders q and m always change with an increase of b (q increasing and m decreasing). For example, after milling of tungsten carbide powder, the specific surface of the powder increasing about eight times, the constants in Eq. (III-5) change as follows: $q_1 = 0.65\ h^{-1}$, $q_2 = 3.91\ h^{-1}$; $m_1 = 28.8$, $m_2 = 3.40$ (sintering in vacuum at 1400°C).

The value of a (numerator of a/b) varies with the type of defect and therefore may differ for powders obtained by different methods [let us remember that a is the pre-exponential factor in Eq. (XI-3) linking the rate of elimination of defects with the concentration of defects].

The value of a also enters into product aN_{in}, which is one of the kinetic constants determining the course of densification. On the basis of a number of considerations (for example, because of the small change in sequence of value of a/b for different powders) it can be assumed that a changes within smaller limits than the concentration of defects, as a consequence of which aN_{in} characterizes the relative concentration of defects in the body at the beginning of isothermal sintering. Nevertheless, there is reason to assume that, despite the sharp reduction in the concentration of defects during heating to sintering temperature, the difference in the concentration of defects in powders of different origins is not eliminated, and the values of aN_{in} found under the same sintering conditions (heating rate and temperature) to some extent also reflect the relative concentration of defects in the original powders. Several observations seem to confirm this.

During annealing (calcining) of powders the change of aN_{in} greatly resembles the recovery of several physical properties. There is a decrease of aN_{in} along with a decrease in the width of x-ray diffraction lines. Evidently the process leading to reduction of aN_{in} coincides with the processes causing the change in the width of diffraction lines (growth of blocks and removal of microstresses). Nevertheless, there is no reason to consider that the small size of the blocks and the large microstresses (second-order stresses) are responsible for the acceleration of flow in defective crystals. Apart from the considerations in this connection given in Chapter VIII, this conclusion also results from the x-ray analysis of the substructure of powders and compacts being sintered. Sharply different values of aN_{in} are observed for nickel powders obtained by reduction of oxides and by dissociation of carbonyl. Since the temperature at which the powder is obtained is lower for dissociation of carbonyl than for reduction of oxides, one would expect the carbonyl powder to have a higher concentration of defects. However, calculations of the value of aN_{in} lead to the opposite conclusion. In sintering with rapid rise of the temperature to 800°C the value of aN_{in} for nickel from oxides was 2.5×10^{10}, while that for nickel from carbonyl was 5×10^5 h^{-1}. X-ray analysis confirmed the higher concentration of defects in the powder obtained from oxides: The width of line (111) was almost 50% larger for the powder obtained from oxides than for the powder from carbonyl. Similar results were obtained by Skorokhod [133], who observed broader (111) and (221) for powder from oxides than for powder from carbonyl. In the given case the concentration of defects calculated from the constants of the kinetic equation agree qualitatively with the values calculated from the x-ray data. However, Skorokhod [133] also obtained data indicating that agreement of the two characteristics is possible only in separate cases and that when powders have defects of different types no such agreement can exist. It was found that the width of the x-ray lines increases sharply after compacting. In this case the width of lines from the carbonyl nickel compact reaches the width of lines from the uncompacted powder obtained by reduction of oxides or even exceeds it. However, such a sharp change in lattice defects is not reflected in the rate of relative reduction in volume of pores, which does not increase with an increase of the compacting pressure.

The absence of any influence of the defects occurring during plastic deformation on the densification rate during sintering and, consequently, on the flow of crystals at small loads is confirmed by several phenomenological characteristics of sintering. As follows from the constant values of the relative reduction in volume of pores for most powders, the rate of reduction in volume of pores per unit volume of pores is independent of the initial density and of the compacting pressure. The degree of deformation of particles, and with it the increase in the concentration of defects induced by compacting, differs for different powders. If the flow rate varied with the concentration of defects occurring during compacting then their effect on densification would be identical during compacting and subsequent sintering of powders of high-ductility and low-ductility substances. Constant values of the relative reduction in volume of pores are observed in sintering of low-ductility (tungsten carbide) and high-ductility powders (copper, silver), which indicates the independence of this factor and, consequently, the independence of the densification process from defects occurring during compacting. The very fact that v_s/v_p is constant with increasing d_p indicates the absence of an accelerating effect of the defects in the lattice induced by deformation on the course of densification: The increase in the concentration of defects with an increase of compacting pressure is not reflected in the relative rate of reduction in volume of pores during sintering. The absence of any relationship between densification and the degree of deformation in compacting was also established in [134] and in the dissertation of Andrievskii.*

Thus, the acceleration of the flow of a crystalline body can be due only to the particular nature of lattice defects occurring during growth of the crystals. X-ray analysis gave a correct estimate of the relative concentration of defects in a crystalline substance that evidently consist of a small percentage of all defects only in case the concentration of defects responsible for acceleration of flow in the powders compared changes in parallel with the total concentration of defects.

Returning now to analysis of a/b, we can state that this constant, equal to $a/\vartheta\beta$, varies with the dispersity of the particles of powder (through the value of β) and with the value of the pre-exponential factors ϑ and a entering into Eqs. (XI-1) and (XI-3), linking the flow rate and the rate of reduction in the concentration of defects with the current concentration of defects. Therefore a/b varies unequivocally with the dispersity only for powders containing certain types of defects for which ϑ and a are constants.

If powders with particles of equal size differ only in the concentration of a given type of defect, then a/b is identical for such powders. This frequently occurs in calcining of powders causing a change in the concentration but not the type of defect. The retention of the shape and slope of curves $v_s/v_p = \varphi(T)$ after calcining of powders is ensured both by constant E_a and by the constant value of a/b.

*R. A. Andrievskii, Dissertation [in Russian], Kiev Polytechnical Institute. Kiev (1959).

The phenomenological interpretation of the constants and comparison of the course of densification with the calculations indicate that Eq. (XI-6) reflects rather completely the kinetics of elementary processes and their interaction during sintering. This equation described the reduction in the volume of pores with time with high accuracy and gives a correct estimate of the variation of densification with temperature, taking account of the limitations imposed by the nonconstant value of αN_{in}.

Chapter XII

Calculating Densification from the Kinetic Constants of the Powder

During isothermal sintering the reduction in volume of pores, as was shown in the preceding chapter, is described by Eq. (XI-6), reflecting the basic characteristics of the kinetics of densification. However, densification during isothermal sintering constitutes only part of the total densification attained in sintering. A substantial portion of the total densification occurs in the period of rising temperature. The rate of reduction in volume of pores at the beginning of isothermal sintering, and the general course of subsequent densification along with it, depends not only on the temperature but also to a considerable extent on the conditions of heating to isothermal sintering temperature. Therefore, for wider use of the kinetic equation (not only for phenomenological analysis of sintering but also for practical purposes) it is necessary to determine quantitatively the changes that occur in the sintered body before isothermal sintering begins.

The proportionality of the rate of densification and the rate of increase in temperature, resulting from the kinetic equation and confirmed by experiment, indicates that at least in some cases the course of densification with a rise in temperature can also be described by an equation of the type of (XI-5). If heating is conducted at a constant rate but the change from rising temperature to isothermal holding is instantaneous then it is possible in principle to calculate the reduction in volume of pores attained in the heating period on condition that the value of aN_0 be previously established, i.e., the kinetic characteristic of the concentration of defects in the original powder. It proves to be very complicated

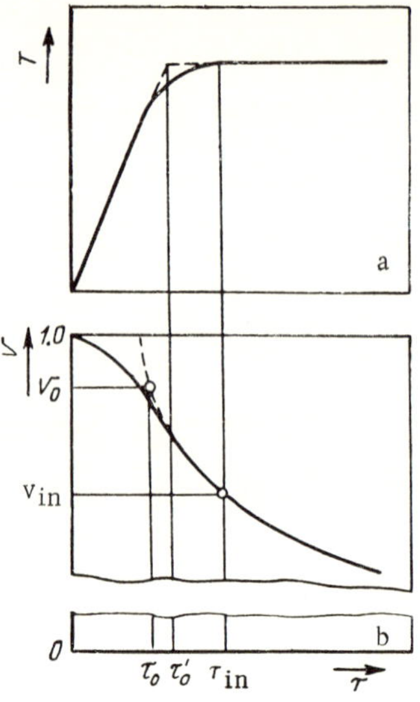

Fig. 38. Variation of T with τ (a) and v with τ (b) for the beginning of sintering (schematic diagram). The dashed lines show the idealized temperature change (a) and the extension of the isothermal curve to v_0 (b).

to realize this possibility, since integration of an equation of the type of (XI-5) with varying temperature cannot be accomplished by means of the integrals of simple functions.

The problem of describing the condition of the sintered body by means of its kinetic characteristics at the time of the change from rising temperature to isothermal holding can be solved by calculations with permissible simplifications, which do, however, produce results that are fairly close to the experimental data. Such calculations are relatively uncomplicated for the case of heating with a constant rate of increase in temperature. Thus, as the basis of the rough calculations we can use the "idealized" heating graph of the sintered body, containing two straight lines, one sloped and one horizontal (Fig. 38a). Under actual sintering conditions such heating is impossible to accomplish, since an instantaneous change from rising temperature to isothermal holding is impossible. However, for real sintering conditions one can plot an idealized graph closest to the actual course of heating. The calculation of densification for such a graph cannot give exact values of densification parameters, although, as practice shows, the difference from the experimental data is not large.*

*Methods of plotting idealized graphs for different sintering conditions will be considered below.

The problem of rough calculation amounts to finding the kinetic characteristics of the sintered body at the time of the change from rising temperature to isothermal holding. The condition of the sintered body, its capacity for further densification, depends mainly on the value of aN, the kinetic characteristic of the concentration of defects. The change in this value with time during isothermal sintering is described by Eq. (XI-15). With even heating at a constant rate of increase in temperature the variation of aN with the temperature reached and the heating rate is expressed by an equation similar to (XI-17):

$$aN = \frac{aN_0}{\frac{aN_0}{\alpha}\int\limits_{T_0}^{T}\exp\left(-\frac{E_a}{RT}\right)dT + 1}, \qquad \text{(XII-1)}$$

where aN_0 is the kinetic characteristic of the concentration of defects in the original powder, E_a is the activation energy of the process of elimination of defects, α is the rate of increase in temperature, deg/h, and T_0 is the temperature before heating begins.

This equation was obtained from (XI-17) by multiplying each side by a. The value of the integral

$$\int\limits_{T_0}^{T}\exp\left(-\frac{E_a}{RT}\right)dT$$

is not taken in integrals of simple functions, although it can be found by numerical integration. For practical purposes it is convenient to use tabular numerical values of the integral.

Integral

$$\int\limits_{T_0}^{T}\exp\left(-\frac{E_a}{RT}\right)dT$$

can be represented as the difference of functions F_T and F_{T_0}, equal to the values of integrals

$$\int\limits_{0}^{T}\exp\left(-\frac{E_a}{RT}\right)dT$$

TABLE 27. Values of log F_T for Values of E_a from 10,000 to 100,000 cal/g-atom

Temperature, °C	log F_T for E_a (cal/g-atom)									
	10,000	20,000	30,000	40,000	50,000	60,000	70,000	80,000	90,000	100,000
500	−0.884	−3.894	−6.874	−9.823	−12.745	−15.648	−18.540	−21.422	−24.298	−27.168
550	−0.654	−3.494	−6.304	−9.082	−11.832	−14.564	−17.284	−19.995	−22.699	−25.398
600	−0.467	−3.147	−5.797	−8.423	−11.021	−13.601	−16.169	−18.728	−21.280	−23.827
650	−0.272	−2.822	−5.342	−7.833	−10.296	−12.740	−15.173	−17.596	−20.013	−22.424
700	−0.101	−2.531	−4.931	−7.301	−9.642	−11.965	−14.276	−16.578	−18.873	−21.163
750	+0.050	−2.270	−4.560	−6.819	−9.050	−11.263	−13.464	−15.657	−17.842	−20.022
800	+0.191	−2.029	−4.219	−6.379	−8.511	−10.625	−12.727	−14.820	−16.906	−18.986
850	+0.323	−1.807	−3.907	−5.977	−8.019	−10.042	−12.053	−14.055	−16.051	−18.041
900	+0.422	−1.618	−3.628	−5.608	−7.566	−9.507	−11.435	−13.355	−15.267	−17.174
950	+0.553	−1.417	−3.357	−5.267	−7.150	−9.014	−10.866	−12.709	−14.546	−16.377
1000	+0.658	−1.242	−3.112	−4.952	−6.764	−8.559	−10.341	−12.114	−13.880	−15.641
1050	+0.771	−1.069	−2.879	−4.659	−6.407	−8.136	−9.853	−11.562	−13.263	−14.959
1100	+0.863	−0.917	−2.667	−4.387	−6.074	−7.744	−9.401	−11.049	−12.690	−14.326
1150	+0.906	−0.803	−2.475	−4.132	−5.764	−7.377	−8.978	−10.571	−12.156	−13.736
1200	+0.985	−0.658	−2.286	−3.890	−5.473	−7.034	−8.584	−10.124	−11.657	−13.185
1250	+1.059	−0.537	−2.117	−3.669	−5.201	−6.713	−8.214	−9.706	−11.190	−12.670
1300	+1.128	−0.447	−1.953	−3.461	−4.945	−6.412	−7.867	−9.313	−10.752	−12.186
1350	+1.194	−0.317	−1.799	−3.264	−4.704	−6.128	−7.540	−8.944	−10.340	−11.731
1400	+1.256	−0.202	−1.642	−3.068	−4.476	−5.861	−7.233	−8.596	−9.952	−11.303
1450	+1.315	−0.102	−1.500	−2.893	−4.262	−5.608	−6.942	−8.267	−9.586	−10.898
1500	+1.372	−0.008	−1.374	−2.727	−4.058	−5.368	−6.667	−7.956	−9.239	−10.516

and

$$\int_0^{T_0} \exp\left(-\frac{E_a}{RT}\right) dT\ .$$

Since F_T is many orders larger than F_{T_0}, the latter function can be neglected, assuming

$$\int_{T_0}^{T} \exp\left(-\frac{E_a}{RT}\right) dT \approx \int_0^{T} \exp\left(-\frac{E_a}{RT}\right) dT = F_T\ ^\circ\mathrm{C}.$$

For comparison with the table it is convenient to use values of $\log F_T$. The difference of $\log F_T$ is almost proportional to the differences in the activation energies, which makes it possible to use proportional interpolation methods.*

The values of $\log F_T$ for activation energies E_a in multiples of 10,000 cal/g-atom are given in Table 27. The values of $\log F_T$ for intermediate values of E_a are found by proportional interpolation. If $E_{a(B)}$ and $E_{a(M)}$ are values (multiples of 10,000 cal/g-atom) closest to the given value of E_a, with $E_{a(B)} > E_a > E_{a(M)}$, and $\log F_{T(B)}$, $\log F_T$, and $\log F_{T(M)}$ are the corresponding values of the logarithms of function F_T, then

$$\log F_T = \frac{\left(\log F_{T(B)} - \log F_{T(M)}\right)\left(E_a - E_{a(M)}\right)}{10{,}000} + \log F_{T(M)}. \qquad \text{(XII-2)}$$

From the logarithm we find F_T, the value of which is used to determine aN_{in}:

$$aN_{in} = \frac{aN_0}{1 + \dfrac{aN_0}{\alpha} \cdot F_T}\ . \qquad \text{(XII-3)}$$

*The value of function F_T can be found also by means of a rough formula proposed by Damask and Dienes [122], using an approximation for the integral of the exponent [135]. With regard to our case the formula will take the form

$$F_T = \frac{RT^2}{E_a} \exp\left(-\frac{E_a}{RT}\right).$$

However, this formula gives fairly precise values of F_T only for activation energies over 25,000-30,000 cal/g-atom.

Thus, there is no difficulty in calculating aN_{in} for given conditions of the beginning of isothermal sintering if the value of aN_0 is known. It is far more complicated to calculate the value of v_{in}, the relative volume of pores at the beginning of isothermal sintering. As already noted, exact calculations are so complicated that they are of no practical importance. The rough calculations given below for $v_s/v_{in} = f(\tau)$ make it possible to plot a curve without determining the value of v_{in}.

Before giving the essentials of the method of calculation we must have some idea of the initial volume of pores at which regular densification begins that can be described by kinetic equations. This volume of pores v_0, which will be expressed by v_s/v_p, is usually not equal to unity. As was indicated for sintering of amorphous bodies (Chapter IX), the regular change in the densification rate due to changing geometry of the pores and expressed by Eq. (IX-1) is observed not at the beginning of sintering but after some reduction of the initial volume of pores to v_0. Let us recall that the regular reduction in volume of pores begins after connected channels are formed. In sintering of crystalline bodies the course of densification at the beginning of heating is altered by individual changes in the shape of the particles due to the elastic-plastic aftereffect, and in the presence of fine pores between particles by rapid reduction in the volume of fine pores. Therefore one of the essential problems of the calculation is to determine the value of v_0 for a given powder. We shall take up the problem of determining v_0 somewhat later.

The rough calculations are based on extrapolation of experimental or theoretical curves to values of v_0 expressed as v_s/v_p. Let us remember that the results of calculating $v_s/v_p = f(\tau)$ do not vary with the choice of point taken as the beginning of the isothermal process. It is important only that this point lie on the isothermal curve. It can be shifted both to longer sintering times and shorter sintering times down to the value of v_0.

Using some modification of Eq. (III-5), one can describe densification beginning from v_0 by the expression

$$v = v_0 (q_0 m\tau + 1)^{-\frac{1}{m}} \qquad \text{(XII-4)}$$

(let us remember that v is the relative volume of pores expressed as v_s/v_p). The curve of this equation coincides with the curve of equation $v = v_{in}(qm\tau + 1)^{-1/m}$ if the values of q_0 and q are linked by the equality

$$q_0 = q\left(\frac{v_0}{v_{in}}\right)^m, \tag{XII-5}$$

where v_0 is the volume of pores at an arbitrary time and v_{in} is the volume of pores at the actual beginning of isothermal sintering. The coordinates of time for these points are, respectively, equal to τ_0 and τ_{in} (see Fig. 38b).

The entire densification is described as isothermal by Eq. (XII-4). Only sections of the curve for $\tau > \tau_{in}$ are real. Sections between τ_0 and τ_{in} are unreal, but the densification after this period is equal to the densification attained in real sintering in the heating period before the beginning of isothermal sintering. Thus, if the constants of Eq. (XII-4) are known then it is possible to calculate densification both in the period between τ_0 and τ_{in}, corresponding to densification with rising temperature, and after τ_{in}, with isothermal sintering.

For practical use of Eq. (XII-4) it is necessary to know constants m and q_0 and also to establish the coordinates of the point corresponding to the origin of the curve of this equation. Constant m is found for a given temperature from a/b and E_a, using Eq. (XI-9). The main problem of the rough calculations will be to determine q_0 from the constants of the powder and also the coordinates of the point from which the curve begins: τ_0 and v_0.

The constants of the powder, from the isothermal sintering data, must also be calculated by (XII-4). By analogy with Eq. (XII-13), the values of aN'_{in} and q_0m are related by the equation

$$aN'_{in} = q_0 m \exp\frac{E_a}{RT}, \tag{XII-6}$$

or, taking into account (XII-5):

$$aN'_{in} = qm\left(\frac{v_0}{v_{in}}\right)^m \exp\frac{E_a}{RT}. \tag{XII-7}$$

Equation (XII-3) could be solved for aN_0 from the experimentally determined value aN'_{in} (from the values of q and m calculated from the experimental variation of v_s/v_p with τ):

$$aN_0 = \frac{aN'_{in}}{1 - \frac{aN'_{in}}{\alpha} \cdot F_T}. \tag{XII-8}$$

However, according to Eq. (XII-7), the value of aN'_{in} varies with v_0. Thus, aN_0 also varies with v_0. With increasing v_0 the value of aN_0 calculated by (XII-8) tends to increase, since aN'_{in} increases with v_0, and the difference $1-(aN'_{in}/\alpha)F_T$ rapidly decreases. At some value of v_0 the expression $(aN'_{in}/\alpha)F_T$ becomes equal to 1 and $aN_0 = \infty$. With further increase of v_0 the calculation becomes senseless, since aN_0 acquires a negative value. All these relationships are conveniently followed on a plot of $1/aN'_{in}$ vs. F_T. Taking (XII-3) in the form

$$\frac{\alpha}{aN'_{in}} = \frac{\alpha}{aN_0} + F_T, \tag{XII-9}$$

we find that the relationship between $1/aN'_{in}$ and F_T is linear. A graph of the relationship from experimental data confirms this (Fig. 39). The linear relationship is retained with aN'_{in} calculated for different values of v_0. Depending on the value of v_0 selected, the straight line will pass above or below the origin of coordinates. The latter case corresponds to an unreal relationship (aN_0 is negative). If the line passes through the origin of coordinates then $aN_0 = \infty$ and Eq. (XII-9) takes the form

$$\frac{\alpha}{aN'_{in}} = F_T. \tag{XII-10}$$

In determining the value of v_0 it must be taken into account that the value of v_0 may change within known limits without detriment to the accuracy of the calculation if τ_0 changes at the same time (we have already noted that in describing densification with kinetic equations any point on the isothermal curve can be chosen as the beginning of the process). Therefore the value of v_0 may be taken as equal to the limit value at which the calculation is still real. This value corresponds to the linear variation of α/aN'_{in} with F_T passing through the origin of coordinates. It is easily found from Eq. (XII-10) and

$$aN'_{in} = aN_{in}\left(\frac{v_0}{v_{in}}\right)^m, \tag{XII-11}$$

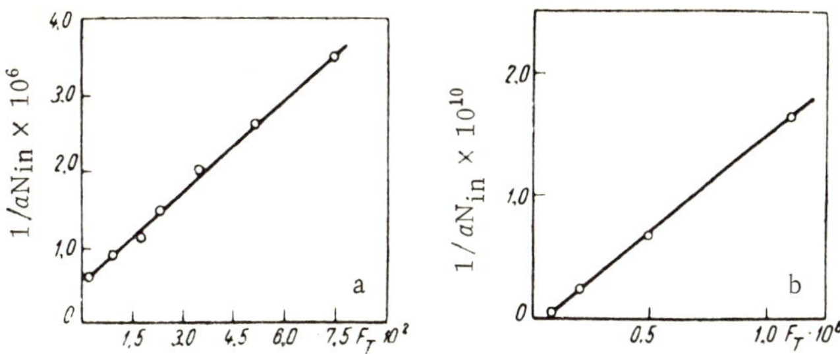

Fig. 39. Plots of $1/aN_{in}$ vs. F_T for nickel (a) and copper powders (b) from experimental data; the values of aN'_{in} are calculated by (XII-7); the constants of the powders are given in Tables 28 and 29.

where aN'_{in} and aN_{in} are the kinetic characteristics of the concentration of defects corresponding to points on the isothermal curve v_0, τ_0 and v_{in}, τ_{in} (see Fig. 38b). This relationship comes from Eqs. (XII-7) and (XI-13). From (XII-10) it follows that in determining the value of aN'_{in} for two isothermal sinterings with v_0 equal to the limit value, when $aN_0 = \infty$, the ratio of values of aN'_{in} is equal to the inverse ratio of the values of F_T:

$$\frac{aN'_{in(1)}}{aN'_{in(2)}} = \frac{F_{T(2)}}{F_{T(1)}}. \qquad \text{(XII-12)}$$

From Eqs. (XII-12) and (XII-11) it follows that

$$\frac{aN_{in(1)}\left(\dfrac{v_0}{v_{in(1)}}\right)^{m_1}}{aN_{in(2)}\left(\dfrac{v_0}{v_{in(2)}}\right)^{m_2}} = \frac{F_{T(2)}}{F_{T(1)}}$$

and

$$v_0 = \left(\frac{v_{in(1)}^{m_1}\, aN_{in(2)}F_{T(2)}}{v_{in(2)}^{m_2}\, aN_{in(1)}F_{T(1)}}\right)^{\frac{1}{m_1-m_2}}. \qquad \text{(XII-13)}$$

Since aN is easily found from constants q and m from the data on isothermal sintering, it is not complicated to determine v_0. Use of the limit value of v_0 greatly simplifies the rough calculation.

Since $aN_0 \gg aN'_{in}$ in this case, one can use in place of (XII-3) the much simpler (XII-10), in which, however, it is necessary to introduce several corrections. From (XII-10) it follows that

$$\frac{aN'_{in} \cdot F_T}{\alpha} = 1.$$

Calculation of this relationship from the experimental data, when aN_{in} is determined from q and m, while aN'_{in} is determined from Eq. (XII-11), gives values of order close to but not equal to unity. The discrepancy is due to the deviation of the values used from the exact relationships determined by kinetic equations (let us remember that in the rough calculations the changes in aN and v with an increase of temperature were calculated separately without taking into account the relationship of the processes causing these changes). Although the calculated ratio differs from unity, it is constant for all temperatures. This results in the fact that $1/aN'_{in}$ vs. F_T is linear and passes through the origin of coordinates. Hence

$$\frac{aN'_{in} \cdot F_T}{\alpha} = K. \qquad \text{(XII-14)}$$

Coefficient K is a correction introduced in connection with the permissible simplifications, and is usually smaller than unity and most often within limits of 0.1–0.9. After v_0 is found from (XII-13), the value of aN'_{in} is determined as $aN_{in}(v_0/v_{in})^m$ and K is calculated from (XII-14).

If K is known then aN'_{in} is easily calculated for any temperature. From (XII-14) it follows that

$$aN'_{in} = \frac{\alpha K}{F_T} \qquad \text{(XII-15)}$$

[let us recall that F_T is a function of temperature and activation energy determined from Table 27 and Eq. (XII-2); α is the average rate of increase in temperature, deg/h].

Thus, when the limit value of v_0 is used it is very simple to calculate aN'_{in} from the constants of the powder.

Let us note that the "disappearance" of aN_0 from the calculations (the relative concentration of defects in the original powder) does not indicate excessive simplification of the calculations. With $aN_0 \gg aN_{in}$ the value of aN_0 has little effect on the value of aN_{in}, which depends mainly on the sintering conditions and not the original value of aN_0. This circumstance is due to the proportionality of the densification rate and the rate of increase in temperature, which, as was noted in Chapter XI, comes from kinetic equation (XII-1), with a large value of aN_0. In this case the units in the denominator can be neglected, since $(aN_0/\alpha)F_T \gg 1$ and the value of aN_0 decreases, i.e., is excluded from the kinetic equation. Inequality $(aN_0/\alpha)\,F_T \gg 1$ is possible only with $aN_0 \gg aN_{in}$. Therefore the proportionality $dv/d\tau \cdot v \sim \alpha$ observed for copper and nickel powders indicates that ordinary powders often have high values of aN_0, ensuring $aN_0 \gg aN_{in}$. It also follows from this that the exact value of v_0 does not differ too greatly from the limit value of v_0.

When low-activity powders are used with a relatively low value of aN_0 the value of α/aN_0 in Eq. (XII-9) cannot be neglected and the value of aN_0 must be taken into account. From a series of isothermal sinterings one can establish only the connection between aN_0 and v_0, but not their values. Therefore, in exact calculations one uses the value of aN_0, which is independent of the isothermal sintering methods. Thus, aN_0 can be determined from the change in the relative rate of reduction in volume of pores $dv/d\tau \cdot v$ with superrapid heating, when the proportionality of $dv/d\tau \cdot v$ and α is disrupted and with increasing α, $dv/d\tau \cdot v$ tends to a level determined by the initial concentration of defects in the original powder. The exact value of v_0 can be found from the value of aN_0 by means of a modification of (XII-8) [Eq. (XII-17), see below], Eq. (XII-7), and constants q and m of the isothermal sinterings of this powder. However, such calculations are needed only for special cases of phenomenological analysis, when the exact values of aN_0 and v_0 must be taken into account. In most cases the densification of low-activity powders can be determined by the rough calculations described above, taking into account the relationship between aN_0 and v_0 with the largest possible values of v_0.

Since, according to (XII-7), aN'_{in} is proportional to v_0^m, for any value aN_{in} one can establish the corresponding value of v_0 at which $(aN'_{in}/\alpha)F_T$ approaches unity, while aN_0, according to (XII-8), tends to infinity. But for low-activity powders this occurs at an unreal high value of v_0. For example, the calculation of v_0 (XII-13) for carbonyl nickel usually gives a value larger than unity (1.3, for example). Although slight expansion of the compact is possible at the beginning of heating, this value of v_0 is excessively high. It must be assumed that in cases where calculations by (XII-13) give v_0 larger than 1 the value of aN_0 is not so high that α/aN_0 can be neglected in (XII-9). For low-activity powders the exact value of v_0 may differ slightly from unity because of the elastic-plastic expansion and because of the possibility of pores between particles, although the deviation from unity is far smaller than for active powders. Therefore, in cases where the calculations give $v_0 > 1$ one must take $v_0 = 1$, but the value of aN_0 must be used in the calculations, which must be found for the given powder. For the case of $v_0 = 1$ the value of aN_0 can be calculated by (XII-8) and (XII-7). In these calculations it is also necessary to introduce a correction factor K, as in calculating the densification of active powders. Let us recall that with a linear variation of $1/aN'_{in}$ with F_T passing through the origin of coordinates one must use in (XII-15), which describes this relationship, the product αK instead of α, where K is the correction factor compensating the inaccuracy caused by the simplifications and assumptions in the rough calculations. In the case where $1/aN'_{in}$ vs. F_T is expressed by a straight line not passing through the origin of coordinates the correction factor K is unnecessary for the same reasons.

Thus, in place of (XII-9) we have

$$\frac{\alpha K}{aN'_{in}} = \frac{\alpha K}{aN_0} + F_T . \qquad \text{(XII-16)}$$

The previously established relationship $\alpha K/aN'_{in} = F_T$ is a particular case of a more general relationship; it is obtained from (XII-16) if $aN_0 \to \infty$. It follows from (XII-16) that a correction factor must also be introduced into Eqs. (XII-3) and (XII-9):

$$aN'_{in} = \frac{aN_0}{1 + \frac{aN_0}{\alpha K} \cdot F_T}, \qquad \text{(XII-17)}$$

$$aN_0 = \frac{aN'_{in}}{1 - \frac{aN'_{in}}{\alpha K} \cdot F_T}. \qquad \text{(XII-18)}$$

If two values of aN'_{in} are known for $v_0 = 1$ calculated from data in two isothermal sinterings at T_1 and T_2, then the value of K is determined by

$$K = \frac{F_{T(1)} - F_{T(2)}}{\alpha\left(\frac{1}{aN'_{in(2)}} - \frac{1}{aN'_{in(1)}}\right)}. \qquad \text{(XII-19)}$$

The value of aN_0 is then calculated from the value found for K, using (XII-18). The kinetic constants of low-activity powders can be found by this means.

Thus, the calculations of the constants of powders and densification for high- and low-activity powders differs somewhat in the step of determining aN'_{in}. The basic principle is the same: The calculations are made for the largest (real) possible value of v_0. For active powders the limiting conditions with maximum v_0 are $1/aN_0 = 0$ and $v_0 < 1$, while for low-activity powders they are $1/aN_0 > 0$ and $v_0 = 1$. The respective kinetic characteristics of the powder with one boundary condition or the other are easily established by means of plotting $1/aN'_{in}$ vs. F_T. If the interpolar line passes above or through the origin of coordinates then the powder is (arbitrarily) a high-activity powder and the constant is calculated by the first method, i.e., v_0 and K are determined. If the interpolated line passes below the origin of coordinates then v_0 is taken as equal to unity and aN_0 and K are calculated (second method of calculation). Plots of $1/aN'_{in}$ vs. F_T for active copper powder and low-activity carbonyl nickel powder are shown in Fig. 39.

Thus, the calculation of constants aN_0, v_0, and K from isothermal sintering data should begin with determination of the value of aN'_{in} for $v_0 = 1$ and the plot of $1/aN'_{in}$ vs. F_T. The value of aN'_{in} is calculated from constants q and m of the isothermal densification curves and by Eq. (XII-7), in which v_0 is taken as equal to unity. The value of function F_T is found by means of Table 27 and the interpolation formula (XII-2).

A line is drawn along the experimental points in the plot of $1/aN'_{in}$ vs. F_T and the method of calculation for the given powder is determined from its position with respect to the origin of coordinates. The same line is used to find the average values of

aN'_{in} used in the calculations. One selects two points on the line not too far from each other with coordinates $aN'_{in(1)}$, $F_{T(1)}$ and $aN'_{in(2)}$, $F_{T(2)}$. If the line passes below the origin of coordinates then the powder is arbitrarily called active. In this case the limit value of v_0 is determined. Taking the logarithm of Eq. (XII-13) for convenience of calculation and replacing $aN_{in}/v_{in}^{m} = aN'_{in}$, which is equivalent to (XII-11), at $v_0 = 1$ we have

$$\log v_0 = \frac{\log aN'_{(2)} - \log aN'_{in(1)} + \log F_{T(2)} - \log F_{T(1)}}{m_1 - m_2}. \qquad \text{(XII-20)}$$

The values of m_1 and m_2 for the same temperatures are determined by Eq. (XI-9), assuming that E_a, E_b, and a/b have already been found from the same series of isothermal sinterings and the experiment with a stepped rise in temperature, as described in Chapter XI.

Factor K for the given powder is determined by (XII-14). Let us remember that for powder of this type $aN_0 \gg aN_{in}$ and $1/aN_0$ is taken as equal to zero.

If the line passes above the origin of coordinates then $1/aN_0 \neq 0$. In that case v_0 is taken as equal to unity, while K and aN_0 are found by Eqs. (XII-19) and (XII-18). Finally, in those rare cases where interpolation of the line on the plot of $1/aN'_{in}$ vs. F_T passes through the origin of coordinates the limit value of v_0 is equal to unity and K is found by Eq. (XII-14) from one of the values of aN'_{in} for the points selected on the line.

These methods are used to find the constants of the powder from which v_s/v_p vs. τ can be calculated for any temperature. The type and shape of the experimental and theoretical curves agree rather well. However, an additional question arises in practice – the coincidence of the curves with time. For this it is necessary to know both the coordinates of the points taken as the arbitrary beginning of the isothermal curve, i.e., it is necessary to know τ_0 along with v_0.

If an already plotted experimental curve of v_s/v_p vs. τ is calculated then τ_0 is found directly from constants q and m and the coordinates of the point taken as the actual beginning of isothermal sintering in calculating q and m (i.e., v_{in} and τ_{in} – see Fig. 38b):

$$\tau_0 = \tau_{in} - \frac{1 - \left(\frac{v_{in}}{v_0}\right)^m}{qm}. \qquad \text{(XII-21)}$$

It is assumed in this case that v_0 has already been found by the method described above.

The matter is more complicated when densification is calculated for sintering with given parameters T, τ, and α. Exact calculations are difficult in this case and the value of τ_0 is determined approximately on the basis of the following considerations.

From the schematic diagram in Fig. 38 it follows that τ_0' (the time at which rising temperature changes to isothermal sintering temperature according to the idealized graph) and τ_0 (the coordinate of the arbitrary beginning of the isothermal curve) do not coincide. Calculations taking account of the difference $\Delta\tau = \tau_0' - \tau_0$, while possible in principle, are fairly complicated. Rough calculations are not necessary, since completely satisfactory results are obtained if τ_0 is taken as equal to τ_0'. Although the calculated and experimental curves differ by the value of $\Delta\tau$, the error in the calculations is small if the isothermal sintering time is not too short. The value of $\Delta\tau$ is usually several hundredths of one hour and in some cases as high as 0.1-0.2 h. Since the densification rate decreases sharply and the slope of the curve of v_s/v_p vs. τ decreases, with increasing τ the difference in v_s/v_p caused by the difference in the curves decreases rapidly. In practice, the use of the rough calculations has shown that with a sintering time of half an hour and longer the difference $\Delta\tau = \tau_0' - \tau_0$ can be neglected with no large error in the calculations. Thus, for

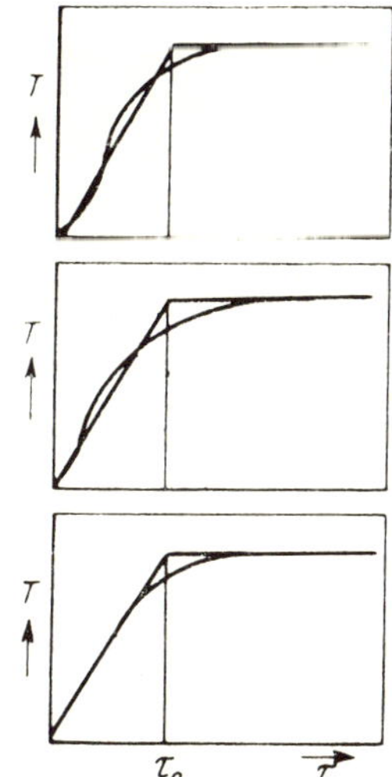

Fig. 40. Methods of constructing idealized graphs with regard to real graphs.

rough calculations the time coordinate can be taken as equal to τ_0', which corresponds to the transition from the sloping line to the horizontal line on the idealized graph.

Two characteristics of the given sintering are found from the idealized graph and used in the calculations: α, the average rate of increase in temperature up to the beginning of isothermal sintering; τ_0, the time of the nominal beginning of isothermal sintering. Thus, a correctly plotted idealized graph from the actual temperature graph of completed or intended sintering is of considerable importance. On the idealized graph the sloped line must be as close as possible to the slope of the section of the curve from the real graph, following the line of the real graph for a fairly long distance. This condition is usually satisfied by a line intersecting the sloped section of the real curve in several places. Figure 40 shows the method of constructing an ideal graph from curves of real graphs that depart from the ideal to a greater or lesser extent.

Together with operating parameters α, τ_0, and T, determining the sintering conditions, and values aN_0, v_0, and K, the methods given above for calculating densification require knowledge of the basic kinetic constants of the powder – the activation energy of the elimination of defects E_a, the activation energy of flow E_b, and the ratio of pre-exponential factors of the kinetic equations a/b.

Methods of determining these constants were given in Chapter XI. Here we point out only that the values of E_a and E_b can be obtained from an experiment with a stepped rise in temperature (with two or more periods of isothermal sintering), and without plotting graphs, directly from the coordinates of the beginning and end of the two adjoining periods of isothermal sintering and the values of m for the same sintering periods. First we find the ratio M:

$$M = \frac{\left[\left(\frac{v_{in(2)}}{v_{fi(2)}}\right)^{m_2} - 1\right]\tau_1}{\left[1 - \left(\frac{v_{fi(1)}}{v_{in(1)}}\right)^{m_1}\right]\tau_2},$$

where $v_{in(1)}$ and $v_{in(2)}$ are the relative volumes of pores at the beginning; $v_{fi(1)}$ and $v_{fi(2)}$ are the relative volumes of pores at the end of the first and second periods of sintering; τ_1 and τ_2 are the times; T_1 and T_2, the temperatures; m_1 and m_2, the constants of Eq. (III-5) for both periods of isothermal sintering (let us recall that the relative volume of pores is determined by the expression $v = v_s/v_p$ for the respective sintering times).

Then we find E_a:

$$E_a = \frac{R \log M}{0.434\left(\frac{1}{T_1} - \frac{1}{T_2}\right)}. \tag{XII-22}$$

The calculation assumes that the change from one temperature stage to the other is completed so rapidly that the concentration of defects does not have time to change substantially and $aN_{in(2)} \approx aN_{fi(1)}$. This equation is easily derived by equating the values of aN for the end of the first sintering period and the beginning of the second period, expressed as v_{in}, v, and m. It is taken into account that $aN_{in} = qm \cdot \exp(E_a/RT)$, while $aN_{fi(1)}$, i.e., the relative concentration of defects at the end of the first period, is equal to

$$\frac{aN_{in(1)}}{aN_{in(1)} \exp\left(-\frac{E_a}{RT}\right)\tau_1 + 1}.$$

The value of E_b is found by adding the values of E_a and ΔE. The values of ΔE and a/b are found from the values of m for a series of isothermal sinterings, as was described in Chapter XI, by Eqs. (XI-11) and (XI-12).

The densification is calculated from the constants of the powder in the following sequence. For the original data and the sintering characteristics of active powders we use constants E_a, E_b, a/b, v_0, and K; for low-activity powders we use E_a, E_b, a/b, aN_0, and K (arbitrary division of powders into high- and low-activity powders is convenient, since it gives a brief characterization of the kinetic type of the powder).

The calculation begins with determining aN'_{in} for given conditions of isothermal sintering (α, τ_0, T). For high-activity powder $aN'_{in} = \alpha K/F_T$; function F_T is found from Table 27 and Eq. (XII-2). Then we find q_0m:

$$q_0 m = aN'_{in} \cdot \exp\left(-\frac{E_a}{RT}\right).$$

Then we determine m:

$$m = \frac{a}{b} \cdot \exp\frac{E_b - E_a}{RT}.$$

The volume of pores is calculated by the equation

$$v = v_0 (q_0 m\tau + 1)^{-\frac{1}{m}},$$

in which we use the calculated values of q_0m and m. The time is counted from τ_0. The total sintering time τ_{tot} is equal to τ_0 + τ. If the total sintering time is known (or given), then the isothermal sintering τ is found by subtracting τ_0 from τ_{tot}. It should be understood that τ_0 does not correspond to the time that true constant temperature is established but to that found from the idealized temperature graph.

For low-activity powder the procedure and calculation are the same except that aN'_{in} is found by

$$aN'_{in} = \frac{aN_0}{1 + \frac{aN_0}{\alpha K} \cdot F_T},$$

and the volume of pores by

$$v = (q_0 m\tau + 1)^{-\frac{1}{m}},$$

since $v_0 = 1$. Let us recall that the calculation is made in relative units of volume of pores, with the unit volume taken as the volume of pores in the original compact. The values of v, v_{in}, and v_0 are equal to v_s/v_p for the corresponding sintering times.

From the values of $v = v_s/v_p$ one can calculate any parameters characterizing the density or porosity of the sintered body (see Chapter XIV).

Despite the use of simplifying assumptions, the rough calculations give completely satisfactory and in some cases unexpectedly high accuracy. As will be shown below, the difference in the calculated and actual values of v_s/v_p is usually only a few percent.

The method of rough calculations has been checked many times, and in all cases the agreement of the calculated and experimental values is completely satisfactory. We present several examples for sintering of a given powder in a wide temperature range. Sintering was conducted in the dilatometric apparatus; densification was observed with isothermal sintering for 2 h. The calculations were made from the constants of the powders. The values of α and τ_0 were determined from the average rate of increase in temperature for a given series of sintering tests and not from the actual values of α and τ_0, so that the calculations corresponded completely to the problem of determining the probable densification from the constants of the powder. The calculated and experimental values of v_s/v_p for 0.5 and 2 h of isothermal sintering for copper, carbonyl nickel, and nickel obtained by reduction of oxides in hydrogen are given in Tables 28-30. The root-mean-square and maximum deviation of the calculated values of v_s/v_p from the experimental values are given in Table 31. The

TABLE 28. Comparison of Experimental and Calculated Values of v_s/v_p for Copper Powder Obtained by Reduction of Oxides in Hydrogen*

Sintering temperature, °C	Value of v_s/v_p for sintering time (h)			
	0.5		2.0	
	Exper.	Calc.	Exper.	Calc.
700	0.87	0.84	0.81	0.80
750	0.77	0.77	0.71	0.72
800	0.67	0.69	0.61	0.62
850	0.56	0.59	0.49	0.51
900	0.49	0.49	0.40	0.39
950	0.37	0.39	0.27	0.29
1000	0.28	0.29	0.20	0.20
1050	0.21	0.21	0.13	0.12

*Constants of powder: E_a = 45,600 cal/g-atom; E_b = 63,000 cal/g-atom; $a/b = 2.62 \times 10^{-3}$; K = 0.51; v_0 = 0.97.

TABLE 29. Comparison of Experimental and Calculated Values of v_s/v_p for Carbonyl Nickel Powder*

Sintering temperature, °C	Value of v_s/v_p for sintering time (h)			
	0.5		2.0	
	Exper.	Calc.	Exper.	Calc.
700	0.83	0.84	0.72	0.74
750	0.75	0.77	0.63	0.64
800	0.68	0.69	0.54	0.54
850	0.59	0.60	0.45	0.45
900	0.51	0.53	0.35	0.37
950	0.44	0.46	0.28	0.30
1000	0.36	0.40	0.22	0.24
1050	0.32	0.34	0.18	0.19

*Constants of powder: E_a = 20,000 cal/g-atom; E_b = 30,100 cal/g-atom; $a/b = 4.39 \times 10^{-2}$; $aN_0 = 1.96 \times 10^5 h^{-1}$; $K = 0.20$; $v_0 = 1$.

accuracy of the calculations is somewhat higher with 2 h of sintering, which is explained by the smaller error in determining τ_0 with increasing isothermal sintering time [see the discussion following Eq. (XII-21)].

When it is considered that the difference between the calculated and experimental values is due not only to the error in the calculations but also random deviations of the sintering conditions from those given (change in the rate of increase in temperature at the beginning of sintering, variation of isothermal sintering temperature) and also errors in determining the density before and after sintering, the agreement observed must be considered better than satisfactory.

With no error in measuring the temperatures and densities and with too much delay in the transition from increasing temperature to isothermal holding, agreement between the calculated and actual values of v_s/v_p is observed continually for all metal powders. Some differences are observed in sintering powders undergoing phase transformations in the process of heating, although with known coordinates of the initial points there is good agreement even in this case. Only densification curves $v_s/v_p = f(\tau)$ at relatively low sintering temperatures are not calculated from the constants of the powder, since the volume of pores at the beginning of isothermal sintering is larger than v_0.

TABLE 30. Comparison of Experimental and Calculated Values of v_s/v_p for Nickel Powder Obtained by Reduction of Oxides in Hydrogen*

Sintering temperature, °C	Value of v_s/v_p for sintering time (h)			
	0.5		2.0	
	Exper.	Calc.	Exper.	Calc.
750	0.60	0.59	0.59	0.57
800	0.54	0.55	0.53	0.52
850	0.47	0.48	0.44	0.44
900	0.39	0.40	0.34	0.33
950	0.26	0.28	0.20	0.20
1000	0.17	0.16	0.11	0.10
1050	0.09	0.07	0.05	0.03

*Constants of powder: E_a = 59,500 cal/g-atom; E_b = 93,100 cal/g-atom; $a/b = 3.91 \times 10^{-6}$; K = 0.16; v_0 = 0.62.

Individual powders have low values of v_0 (nickel, for example), for which results are given in Table 30. A low value of v_0 is usually due to the spongy structure of particles, with a large number of fine pores within the particles, or to a powder in which the size of the particles differs, with a large percentage of fine particles. A high rate of reduction in the number of fine pores leads (with $v > v_0$) to acceleration of densification and to deviation of the course of densification from the law determined by the kinetic equation.

TABLE 31. Root-Mean-Square and Maximum Deviation* of Calculated from Experimental Values of v_s/v_p

Powder	Isothermal sintering time, h	Number of tests	Root-mean-square deviation	Maximum deviation
Copper	0.5	8	0.017	0.03
	2.0	8	0.013	0.02
Carbonyl nickel	0.5	8	0.020	0.04
	2.0	8	0.012	0,02
Reduced nickel	0.5	7	0.011	0.02
	2.0	7	0.012	0.02

*Deviation expressed in percent v_p (v_p taken as unity).

With a relatively small change of v_s/v_p under these sintering conditions (low temperature and $v > v_0$) the calculation of constants q and m in Eq. (III-5) from the experimental data is still possible. In this case the values of q and m will depart from the general law corresponding to normal kinetics of densification. This can be traced from the change in the constant m for isothermal sintering of nickel powder obtained by reduction of oxides in hydrogen. As follows from Eq. (XI-9), the value of m decreases regularly with increasing temperature. With fairly high sintering temperatures for powder in which the particle size differs greatly and the particles are spongy the presence of pores within particles or fine grains does not affect relationship $m = f(\tau)$, since fine pores and sections of fine grains disappear before the beginning of isothermal sintering. At high temperatures one observed a normal variation of m with temperature, while at low temperatures and small total densification the value of m again decreases.

Temperature, °C . . .	1050	1000	950	900	850
Value of m	2.55	2.98	4.11	5.79	16.18
Final value v_s/v_p	0.05	0.11	0.20	0.33	0.44
Temperature, °C . .	800	750	700	650	600
Value of m	23.61	65.6	82.3	44.5	40.5
Final value v_s/v_p	0.53	0.59	0.66	0.74	0.83

This variation of m indicates a substantial increase in the effect of the geometric factor (or an increase of the geometric component of the densification rate) as compared with relationship $dv/d\tau \sim v$. The increase of the geometric component is due to the rapid reduction of fine pores in conformity with the phenomenological interpretation of b (see Chapter XI), induces its increase, and, consequently, m decreases. Thus, the phenomenological importance of v_0 is confirmed by the deviation of the change in m from the general law with $v > v_0$.

The densification can be calculated from the constants of the powder for more complex cases of sintering – for example, for several sinterings with successively increasing temperatures. In this case the temperature of the sintered body between periods of the isothermal sintering may decrease to room temperature or after completion of one period may rise to the temperature of the following period (sintering with a stepped rise in temperature). The calculations give values of v_s/v_p that are quite accurate for practical purposes and, from them, the values of d_s or P_s for any stage of sintering with a stepped rise in temperature. Let us demonstrate this with the previously described experiments (Chapter XI) employing copper, silver, and nickel powders. The curve in Fig. 30a reflecting the change in v_s/v_p with stepped heating of a copper compact can be calculated from the constants of the powder: E_a = 48,400 cal/g-atom; E_b = 62,700 cal/g-atom; $a/b = 1.095 \times 10^{-2}$; $v_0 = 0.94$; K = 0.22. The calculated value of v_s/v_p at the end of the first period of isothermal sintering at 700°C is 0.78, which almost coincides with the experimental value 0.77. Then, calculating aN_{in} for the second period of isothermal sintering by Eq. (XI-15), we find v_s/v_p for the end of the second temperature stage by means of Eqs. (XI-8), (XI-9), and (III-5); the calculations are made in the same sequence for the third period of sintering. The curves obtained are shown in Fig. 30a by the solid lines.

The same curves match the calculations given in Chapter XI. The curves calculated from the experimentally established characteristics of the first sintering period (the method of calculation used in Chapter XI) and curves completely calculated from the constants of the powders only by the method described coincide. In the first case the following stages of sintering are also calculated from the constants of the powder, and therefore when the theoretical and experimental curves coincide for the first stage of sintering they also coincide for the following stage, since the method of calculation is the same.

This same method can be used to calculate the densification of nickel and silver powders with a stepped rise in temperature. The curves shown in Fig. 30b, c were completely calculated from the constants of the powder (for nickel E_a = 72,000 cal/g-atom; E_b = 95,000 cal/g-atom; $a/b = 8.06 \times 10^{-4}$; $v_0 = 0.73$; K = 0.78; and for silver: E_a = 54,000 cal/g-atom; E_b = 73,000 cal/g-atom; $a/b = 1.15 \times 10^{-4}$; $v_0 = 0.82$; K = 0.46).

With a rapid change of temperature from the first to the second sintering period (in 3-5 min) the change in the concentration of defects can be neglected and the concentration of defects $aN_{in(2)}$ for the beginning of the second period can be taken as equal to the concentration $aN_{fi(1)}$ at the end of the first period of isothermal sintering. However, if the temperature is changed slowly then the change in the concentration of defects can be taken into account by calculating $aN_{in(2)}$ from the value of $aN_{fi(1)}$ and α, the rate of increase in temperature between T_1 and T_2. The calculation is made by Eq. (XII-3), in which $\Delta F_T = F_{T(2)} - F_{T(1)}$ is substituted for F_T and aN_0 is replaced with $aN_{fi(1)}$, the relative concentration of defects at the end of the first sintering period. The latter is found by Eq. (XI-15), in which we use the value of aN'_{in} determined by the usual means from the constants of the powder and the values of α and T for the first sintering period.

When several isothermal sintering periods do not follow immediately one after the other and the sintered body is cooled to room temperature the calculation is made in the same sequence as in the experiment with a stepped rise in temperature. In this case the value of aN_0 for the following sintering period is taken as equal to $aN_{fi(1)}$ at the end of the preceding sintering period. In other respects the calculation is made from the kinetic constants of the powder, taking account of its value of α. In all subsequent sintering periods the value of v_0 is taken as equal to unity (for the first sintering period the calculation is made on the general principle that v_0 be equal or not equal to unity). The total densification after several sintering periods will be determined by the values of v_s/v_p, which is equal to the product of the values of v_s/v_p for the individual sintering periods.

A qualitative explanation for the effect of low-temperature sintering on densification with subsequent sintering at higher temperature was given in Chapter XI. Calculation of the changes occurring in the sintered body during the initial heating period and preliminary sintering makes it possible to determine this effect quantitatively. By means of the calculations described above it is not difficult to determine to what extent presintering at a given temperature reduces densification during subsequent high-temperature sintering, and after both sinterings. The experimental and calculated values of v_s/v_p for copper and nickel powders sintered under different conditions are given in Table 32.

As already noted, presintering does not always reduce densification so much during resintering that the total densification is less than the densification resulting from single sintering at high temperature. If the presintering temperature is not far different from that of the following high-temperature sintering then

TABLE 32. Values of v_s/v_p for Copper and Nickel Powders With Sintering Under Different Conditions

Sintering conditions	Value of v_s/v_p					
	Copper*		Nickel, reduced†		Nickel, carbonyl‡	
	Exper.	Calc.	Exper.	Calc.	Exper.	Calc.
Single, 750°C	0.70	0.72	0.55	0.57	0.61	0.64
Double, 750 and 900°C	0.48	0.45	0.41	0.40	0.33	0.31
Single, 900°C	0.40	0.39	0.30	0.33	0.39	0.37

*Constants of powder: E_a = 45,600 cal/g-atom; E_b = 63,000 cal/g-atom; $a/b = 2.62 \times 10^{-3}$; K = 0.51; v_0 = 0.97.
†E_a = 59,500 cal/g-atom; E_b = 93,100 cal/g-atom; $a/b = 3.91 \times 10^{-6}$; K = 0.16; v_0 = 0.62.
‡E_a = 20,000 cal/g-atom; E_b = 30,100 cal/g-atom; $a/b = 4.39 \times 10^{-2}$; $aN_0 = 1.96 \times 10^5$ h^{-1}; K = 0.20.

considerable densification occurs in presintering and the total densification is larger than densification after single high-temperature sintering. Also, the degree of reduction in the densification rate after presintering depends on the properties of the powder. For dispersed powders with relatively low aN_0 the effect of presintering is less pronounced. For such powders presintering does not induce a sharp reduction of densification and sintering twice results in larger densification than single sintering even with a large difference in temperature between the first and final sintering periods. This can be observed in the case of carbonyl nickel powder (see Table 32).

Thus, calculations from the constants of the powder make it possible to determine the conditions in which sintering twice results in less densification than sintering once. The phenomenological analysis, based on the variation of flow with the concentration of defects, not only explains the reduction in the capacity for densification after annealing of a crystalline body at low temperatures but also permits quantitative calculation of this effect by means of the kinetic equations.

Quantitative calculations of the changes occurring in the sintered body before the beginning of isothermal sintering also make it possible to determine with fair accuracy from the con-

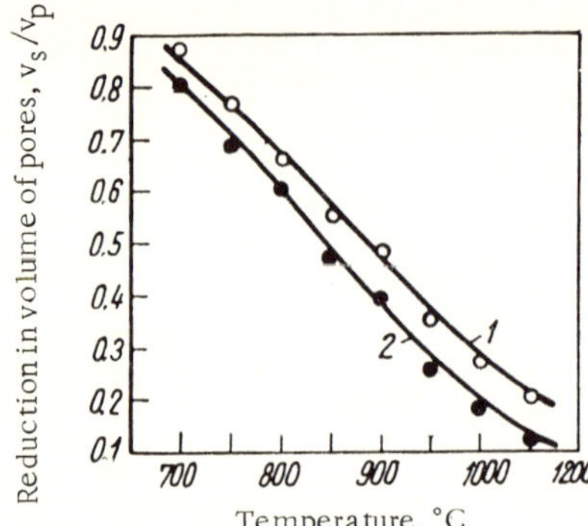

Fig. 41. Calculated (solid lines) and experimental (points) variation of v_s/v_p with temperature for copper powder reduced from oxides at different isothermal sintering times. 1) 0.5 h; 2) 2 h.

stants of the powder the variation of v_s/v_p with temperature at a constant isothermal sintering time and a given rate of increase in temperature before the beginning of isothermal sintering. As was shown in Chapter XI, the calculation of this relationship for the hypothetical case of instantaneous heating to isothermal sintering temperature with constant values of aN_{in} at all temperatures gives on the plot of v_s/v_p vs. T a curve differing substantially from the experimental curve. With use of the method of calculation described, taking into account the change in aN in the initial heating period, this relationship is characterized by a curve that is very close to the experimental curve (Fig. 41 for copper powder and Fig. 42 for nickel powder).

Thus, the method of rough calculation makes it possible, without resorting to sintering tests, to determine the probable degree of densification of compacts from a given powder under different sintering conditions. The basic relationships linking the

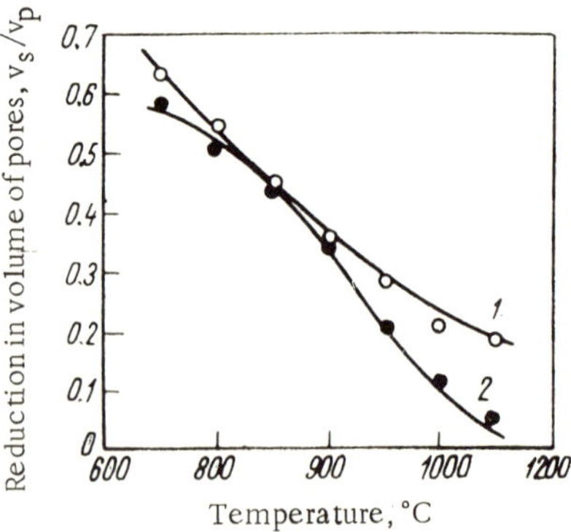

Fig. 42. Calculated (solid time) and experimental (points) variation of v_s/v_p with temperature for nickel powder obtained by different methods and sintered at constant temperature 2 h. 1) Carbonyl nickel; 2) nickel obtained by reduction of oxides in hydrogen.

densification attained with the sintering conditions can also be calculated. The kinetic constants of the powder completely determine the characteristics and form of these relationships.

It should be kept in mind that the calculations give actual values of densification parameters only under conditions of undisturbed densification, i.e., with initial constant density v_s/v_p (or little change). Therefore it is expedient to determine first the "critical values" of density above which v_s/v_p begins to increase (see Chapter II). In some cases (carbonyl nickel powder, powders of stainless and other steels, carbide powders) the critical values of density are not reached even at high compacting pressures, and therefore the calculations for such powders are possible for all compacting pressures and initial densities encountered in practice.

From the value of v_s/v_p other densification parameters can also be calculated – the density and porosity after sintering, shrinkage. The methods of calculating these values are given in Chapter XIV.

Chapter XIII

Clarification of the Nature of Phenomenologically Elementary Processes and Unresolved Problems of Theory

The phenomenological analysis of elementary processes that occur during sintering leads to a number of conclusions on the lattice defects responsible for accelerated flow of a crystalline substance:

1. These defects constitute a small fraction of the defects in real crystals.
2. The defects responsible for accelerated flow may differ substantially in their energetic characteristics for different powders. The activation energy of the elimination of defects may vary within wide limits, depending on the conditions in which the powder was obtained.
3. The activation energy of the elimination of defects is independent of the concentration of defects. It is a constant for a powder obtained by a given method.
4. The process of the elimination of defects responsible for accelerated flow obeys second-order kinetics.

These statements reflect the existing relationships between processes that occur during sintering and the properties of the sintered substance. The last two statements make it possible to consider the elimination of defects as a phenomenologically elementary process (let us recall that the characteristic feature of a phenomenologically elementary process is conformity of the process with a simple kinetic law, defined as the order of reaction, and the presence of a temperature dependence described by the constant value of the activation energy).

To a certain extent it was unexpected that the variation of flow with time obeys a general law reflecting the same kinetics of the elimination of defects for crystalline substances with quite different properties. This unity is all the more remarkable in that the recovery of other properties varying with the concentration of defects obeys quite different laws in the case of different crystalline substances, and sometimes there is no law that can be expressed by a simple kinetic equation. Nevertheless, the unity of the laws of densification, and consequently of the elimination of defects, for substances with such different properites as copper and tungsten carbide is now firmly established fact.

The possibility of considering the elimination of defects as a phenomenologically elementary process is very convenient for analyzing the course of densification. In particular, it permits calculation of densification by means of the constants characterizing the defects in a given powder. In spite of that, phenomenological analysis takes us only so far toward an understanding of the physical nature of the elementary processes.

A phenomenologically elementary process may be a complex process in its physical nature. Under certain conditions a combination of several elementary atomic-molecular processes may appear to be a phenomenologically elementary process. If the change in the properties during annealing of imperfect crystals is due to the development of several concurrent irreversible processes, then when the properties vary with the original effects the time and temperature dependences of the observed change in the properties will reflect the kinetics of the first process. With a succession of several reversible processes the kinetics of the change in the properties will be determined by the kinetics of the slowest process. However, the development of reversible processes is less probable in a defective lattice.

Another conclusion results in the case where the property investigated varies not with the initial concentration of "activated complexes" (defects) entering into the reaction, but with the concentration of intermediate "reaction products."

Let us consider a series of transformations $A_c \rightarrow B_c \rightarrow D_c$, where A_c, B_c, and D_c are atomic complexes capable of passing into the activated condition (lattice defects, defects of molecules, etc.). For simplicity let us assume that all the reactions are irreversible and are first-order reactions: $A_c \rightarrow \delta B_c$ and $B_c \rightarrow \xi D_c$. If the concentration of complexes A_c and B_c is equal to C_A and C_B, then

$$\frac{dC_A}{d\tau} = -\rho C_A, \tag{XIII-1}$$

$$\frac{dC_B}{d\tau} = -\omega C_B, \tag{XIII-2}$$

with

$$\rho = p \cdot \exp\left(-\frac{E_\rho}{RT}\right) \tag{XIII-3}$$

and

$$\omega = k \cdot \exp\left(-\frac{E_\omega}{RT}\right). \tag{XIII-4}$$

If the property varies with the concentration of C_A then the changes in the property will be determined completely by the kinetics of the first reaction. It is a different matter if the property varies with the concentration of C_B. Let us consider this case. If $dC_A/d\tau \ll dC_B/d\tau$ (with $C_A = C_B$) then concentration C_B is determined by the quasiequilibrium conditions corresponding to equality of complexes B_c that are formed and that disappear. From the equality

$$\delta \frac{dC_A}{d\tau} = \frac{dC_B}{d\tau}$$

it follows that the concentration of C_B will be a simple function of C_A

$$\delta\rho C_A = \omega C_B$$

or

$$C_B = \frac{\delta\rho}{\omega} C_A. \tag{XIII-5}$$

Since a change of the concentration of C_B will follow from a change in the concentration of C_A, in this case the change in the property will be determined by the kinetics of the first reaction, but the apparent activation energy of the process found from the temperature dependence of the change in the property will not equal the activation energy of the first reaction. The relationship of the concentrations of C_A and C_B will depend on the temperature: It follows from Eqs. (XIII-5), (XIII-3), and (XIII-4) that for a given C_A the concentration of C_B will be proportional to $\exp[-(E_\rho - E_\omega)/RT]$, i.e., it will increase exponentially with temperature. Thus, the kinetic equation should have an additional exponent, the value of which increases exponentially with temperature. When there is a simple relationship between the property and the concentration of C_B the apparent activation energy of the set of processes inducing the change in the given property will be the value of a constant not depending on the extent of recovery or temperature. It will have features of both a phenomenologically elementary process (constant activation energy and kinetics determined by the order of reaction) and a complex multistage process. The distinguishing feature of this case is that the apparent activation energy of the process is not equal to the activation energy of one of the physically elementary processes but is a more or less complex function of the activation energy of several processes.

It can be assumed that what has been said also applies to the activation energy of flow, determined by the method described in Chapter XI. It is probable that the increase of the flow is due not to the original lattice defects formed in the growth of the crystal but to particular defects ("active" or "mobile" defects) formed in the interaction and annihilation of the original defects. These defects in turn interact with others and disappear at a high rate. The activation energy of the interaction of active defects is far lower than the activation energy of annihilation or interaction of original stable defects. Let us consider several possibilities that result from such an interpretation of the process of elimination of defects.

Since the kinetics of the entire process is determined by the order of the first slow reaction, while the second order is experimentally established for the whole process, the transformation of stable (original) defects into active defects must be ascribed to second-order kinetics. If the concentration of stable defects is equal to N, then

$$\frac{dN}{d\tau} = -\rho N^2. \qquad \text{(XIII-6)}$$

Let us assume that with the disappearance of one stable defect there are formed δ active defects. Then the formation rate of active defects is equal to

$$\frac{dM}{d\tau} = -\delta\frac{dN}{d\tau} = \delta\rho N^2.$$

If second-order kinetics is also assumed for this process, then the rate of disappearance of active defects is equal to

$$\frac{dM}{d\tau} = -\omega M^2. \qquad \text{(XIII-7)}$$

According to the condition of quasiequilibrium (considering that $dN/d\tau \cdot N \ll dM/d\tau \cdot M$), the rates of the disappearance and formation of active defects is equal in first approximation to

$$\omega M^2 = \delta\rho N^2$$

and

$$M = \left(\frac{\delta\rho}{\omega}\right)^{\frac{1}{2}} N. \qquad \text{(XIII-8)}$$

The change in the concentration of stable defects with time is determined by the equation obtained by integration of Eq. (XIII-6):

$$N = \frac{N_{in}}{\rho N_{in}\tau + 1},$$

where N_{in} is the initial concentration of stable defects. Combining the last two equations, we obtain an equation describing the change in the concentration of active defects:

$$M = \frac{\left(\frac{\delta\rho}{\omega}\right)^{\frac{1}{2}} N_{in}}{\rho N_{in}\tau + 1}. \qquad \text{(XIII-9)}$$

The variation of the flow rate with the concentration of defects, and with it the relative rate of reduction in volume of pores, will be expressed by an equality similar to Eq. (XI-2):

$$\frac{dv}{d\tau} = -B_M M v, \qquad \text{(XIII-10)}$$

where $B_M = b_M \cdot \exp(-E_b/RT)$. Coefficient b_M has the same meaning as b in Eq. (XI-2), but with regard to active defects. Combining (XIII-10) and (XIII-9), we have

$$\frac{dv}{v} = -\frac{B_M\left(\frac{\delta\rho}{\omega}\right)^{\frac{1}{2}}N_{in}}{\rho N_{in}\tau + 1}\cdot d\tau \qquad \text{(XIII-11)}$$

and

$$\ln\frac{v}{v_{in}} = -\frac{B_M\delta^{\frac{1}{2}}}{\rho^{\frac{1}{2}}\omega^{\frac{1}{2}}}\cdot\ln(\rho N_{in}\tau + 1). \qquad \text{(XIII-12)}$$

Thus, the derivation of a kinetic law for the formation of intermediate active defects gives an equation equivalent to that obtained earlier for the direct influence of the original defects on flow – Eq. (XI-6). Either equation can be used to calculate densification from previously found constants of the powders, although the new interpretation of the process of elimination of defects has some advantages. Using the same method as in deriving relationships (XI-7) and (XI-9), let us find the value of the constants of Eq. (III-5) expressed as the constants of Eq. (XIII-12):

$$m = \frac{\rho^{\frac{1}{2}}\omega^{\frac{1}{2}}}{B_{in}\delta^{\frac{1}{2}}} \quad \text{and} \quad q = \delta^{\frac{1}{2}}\left(\frac{\rho}{\omega}\right)^{\frac{1}{2}} B_M N_{in}.$$

The values of ρ, ω, and B_M vary exponentially with temperature [(XIII-3), (XIII-4), (XIII-10)] and therefore the variation of q with temperature will have the form

$$q = \left(\frac{dv}{d\tau\cdot v}\right)_{in} = \vartheta\exp\left[-\frac{E_b + \frac{1}{2}(E_\rho - E_\omega)}{RT}\right]N_{in}, \qquad \text{(XIII-13)}$$

where $\vartheta = (\delta\rho/k)^{1/2}\cdot b_M$.

Let us recall that q is equal to the rate of relative reduction in volume of pores at a concentration of defects equal to N_{in}. It was previously shown that except for the inhibiting effect of a varying concentration of defects (the original "stable" defects) the values of the activation energies found from the temperature dependence $dv/d\tau \cdot v$ may have very high values that exceed the self-diffusion activation energy. The temperature dependence of the flow rate, which can be seen from (XIII-11) and is reflected by (XIII-13), gives an explanation of this. Since $E_\rho > E_\omega$, the apparent activation energy exceeds the "true" activation energy of flow with a constant concentration of active defects. The additional term in the exponent characterizes the change in the quasiequilibrium concentration of active defects with an increase of temperature.

Thus, the mechanism of the elimination of defects with formation of intermediate active defects satisfactorily explains the sharp increase in the flow rate with an increase of temperature: Together with the acceleration of flow, which would also occur with a constant concentration of defects, the number of active defects increases, which additionally increases the flow rate.

Second-order kinetics was assumed in the derivation of Eq. (XIII-12) to describe the interaction and annihilation of active defects. However, it is possible to arrive at the same equation of the kinetics of densification if another (actually any) order of reaction is assumed, on condition that a corresponding change occur in the form of the variation of flow with the concentration of defects. It is possible, for example, to assume that the common prerequisite both for flow at small loads and for the interaction and annihilation of defects is a special "unstable" condition of the crystal lattice. If it is further assumed that the degree of "instability" increases in proportion to N^n then the kinetic equations of these processes will take the form

$$\frac{dv}{d\tau \cdot v} = -BN^n \qquad \text{(XIII-14)}$$

and

$$\frac{dN}{d\tau \cdot N} = -AN^n. \qquad \text{(XIII-15)}$$

Let us note that the latter expression establishes the connection between the rate of disappearance of defects, in terms of a single defect or per unit concentration of defects, with the total concentration of defects. It indicates that the probability of the disappearance of each defect depends on the condition of the lattice surrounding it, expressed by the value N^n.

From (XIII-15) it follows that $N = N_{in}(nAN_{in}^{n}\tau + 1)^{-1/n}$. Substituting this expression into (XIII-14), and after integrating, we obtain

$$\ln\frac{v}{v_{in}} = -\frac{B}{nA}\ln\left(nAN_{in}^{n}\tau + 1\right). \qquad \text{(XIII-16)}$$

This equation is completely analogous to (XIII-12) or (XI-6).

Thus, the assumption that the elimination of defects obeys second-order kinetics is not an absolute necessity for phenomenological analysis of densification. The kinetic equation of densification can be arrived at if it is assumed that both flow and the disappearance of defects vary with the instability of the lattice, which is a power function of the concentration of defects. However, phenomenological analysis is simplified if it is taken that $n = 1$ in (XIII-14) and (XIII-15), which corresponds to second-order kinetics (for elimination of defects). The estimate of the behavior of a sintered body under different sintering conditions does not change because of this. At the same time, the assumption of the complex variation of flow with the concentration of defects that is more complicated than a simple proportionality has some basis in fact. With direct proportionality the enormous differences in flow would compel the assumption that there is a superhigh concentration of defects, which is hardly possible.

The author by no means insists on the reality of the mechanism considered above for the connection between the kinetics of densification and lattice defects with use of the concept of the formation of intermediate active defects. It is quite possible that a more thorough analysis based on additional experimental data and modern concepts of metal physics will permit a more valid interpretation of this relationship. The discussion is presented only to show the possibility of the existence of features of a phenomenologically elementary process in complex physically elementary processes. Nevertheless, the analysis above can provide further explanation of the quite different values of the activation energy of flow with different methods of obtaining crystalline substances with a defective lattice.

The phenomenological analysis shows that the temperature dependence of densification is determined by constant values of the activation energy of elementary processes. By means of the kinetic equation including the constant value of the activation energy it is possible to describe or explain the basic features of densification associated with a change in temperature (the course of densification in single or multistage sintering, with a stepped rise in temperature, the effect of the rate of increase in temperature on densification, etc.). However, identifying the physical nature of phenomenologically elementary processes can substantially correct these concepts. It is not excluded that the constant activation energy of the elimination of defects is only an apparent constant. Some combinations of physically elementary processes, including processes with variable activation energies, may as a whole obey such a temperature dependence that appears to correspond to a constant activation energy. As the simplest example one can cite the linear variation of the activation energy of the elimination of defects of radiation origin with temperature that has been found by several investigators [136]. This variation can be conceived as the result of the disappearance at a given temperature of defects with a relatively low activation energy. Raising the temperature "brings into play" defects with a higher activation energy, as a consequence of which the activation energy of the most active defects increases with temperature (linear in the simplest case). If $E = E_0 + \Omega T$ then the exponential term in the kinetic equation is equal to

$$\exp\left(-\frac{E_0+\Omega T}{RT}\right)=\exp\left(-\frac{E_0}{RT}\right)\exp\left(-\frac{\Omega}{R}\right),$$

and the temperature dependence will have the same form as in the case where the process depends on a constant value of the activation energy, equal to E_0. For the defects responsible for acceleration of flow there is little basis for this hypothesis, although other possibilities exist for creating conditions satisfying the apparent constant activation energy. We shall not consider these possibilities, since for our case they are too problematical. Valid hypotheses must be based either on the mechanism of the elimination of defects in the recovery of some property or other (on condition of evidence that these defects are responsible for acceleration of flow) or on experimental data that would show the complex nature of phenomenologically elementary processes developing during sintering. However, the arrangement of an experiment capable of revealing the complex character of these processes is a very complicated problem. It is difficult to find conditions in which the process of densification would deviate from the course determined by the development of phenomenologically elementary processes. This complicates the problem of revealing the true nature of the elementary processes to an extreme degree. But this same circumstance also facilitates matters somewhat: It ensures the possibility of wide use of simple kinetics of elementary processes to analyze and explain many features of the densification process during sintering at the level of the phenomenological theory.

The phenomenological theory leaves unresolved a number of questions concerning the mechanism of sintering. Phenomenological analysis is based on the concept of the variation of flow with the concentration of defects. The proportionality of the flow rate and the concentration of defects or some function of the concentration of defects assume the absence of flow of the crystal not containing defects responsible for the acceleration of flow. This concept is quite valid at relatively low temperatures. But at high temperatures close to the melting point it is necessary to consider the possibility that another flow mechanism develops, which may be diffusion-viscous flow according to Nabarro-Herring. Judging from the kinetic constants of diffusion creep of metal wire under loads close to those created by surface tension [137-142], the participation of this mechanism seems probable in sintering of metal powders at temperatures of $0.7-1.0T_m$. If the diffusion mechanism coexists with a special mechanism associated with lattice defects then it can be assumed that the effect of the diffusion mechanism, masking the second mechanism with a high concentration of defects, may be notable at the low concentration of defects established after prolonged sintering. Experiments with prolonged sintering of different metal powders were made in this connection. It was assumed that after considerable reduction of the densification rate because of the reduction in the concentration of defects the effect of the diffusion component in the flow rate would become notable, leading to deviation of the reduction in volume of pores from the curve corresponding to Eq. (III-5). Unfortunately, it was not possible to conduct sintering by the same method for the projected period of 500 h. The experiment was made as follows. The initial sections of the isothermal curves of $v_s/v_p = f(\tau)$ were obtained by sintering of compacts in the dilatometric apparatus. Then the sintered samples were transferred to an ordinary tubular furnace, where sintering was conducted at the same temperature

TABLE 33. Variation of v_s/v_p with τ during Prolonged Sintering of Copper, Carbonyl Nickel, Iron, and Cobalt*

Isothermal sintering time, τ, h	Value of v_s/v_p							
	Copper (q = 1.19 h^{-1}, m = 10.30)		Nickel (q = 0.63 h^{-1}, m = 2.30)		Iron (q = 3.99 h^{-1}, m = 12.27)		Cobalt (q = 0.84 h^{-1}, m = 13.70)	
	Exper.	Calc.	Exper.	Calc.	Exper.	Calc.	Exper.	Calc.
0	0.800	0.800	0.646	0.646	0.653	0.653	0.546	0.546 (P. T.)
0.5	0.661	0.561	0.560	0.560	0.497	0.497	0.475	0.475 (P. T.)
2	0.584	0.584	0.358	0.358	0.445	0.445	0.433	0.433 (P. T.)
3.9	0.545	0.549	0.280	0.285	0.422	0.422	0.408	0.413
6	0.525	0.528	0.239	0.244	0.411	0.408	0.395	0.401
20	0.504	0.468	0.151	0.148	0.374	0.369	0.363	0.365
74	0.433	0.413	0.081	0.085	0.352	0.332	0.350	0.338
110	0.423	0.407	0.073	0.071	0.347	0.321	0.350	0.324
170	0.391	0.381	0.054	0.059	0.345	0.319	0.342	0.314
320	0.360	0.357	0.045	0.046	0.334	0.295	0.295	0.300
500	0.343	0.340	0.037	0.038	0.329	0.284	0.273	0.290

*Sintering temperature 800°C.

with periodic interruptions for cooling and measurement of the dimensions. Sintering was interrupted every 16, 24, and then 48 h. The experimental conditions were not very strict, although the results are still of interest.

The temperature controls of the dilatometric furnace and the tubular furnace were identical (automatic temperature control ensuring constant temperature within limits of ±5°C). The change in the volume of pores, expressed as v_s/v_p, during prolonged sintering of copper, nickel, cobalt, and iron compacts is given in Table 33.

The experimental values of v_s/v_p differ little from the calulated values* obtained by extrapolation with use of Eq. (III-5). For copper the intermediate values of v_s/v_p are somewhat higher than the calculated values, although at the end of sintering they again approach the calculated values. For iron the values are somewhat higher than the calculated values, and remain so up to the end of sintering. The values for cobalt are also somewhat higher, but at the end of sintering the densification unexpectedly accelerates. For carbonyl nickel there is good agreement between the experimental and calculated values of v_s/v_p despite an extremely low residual porosity at the end of sintering (~5%). With insufficiently strict experimental conditions (there was no guaranteee of complete coincidence of the temperature in the zones of both furnaces where the samples were placed) the slight deviation from the course of densification determined by Eq. (III-5) is difficult to ascribe to the effect of the diffusion mechanism coexisting with the mechanism associated with the effect of defects. On the whole, no very notable effect of an additional mechanism of densification was observed in prolonged sintering.

Further research is needed on the question of high-temperature and low-temperature mechanisms of sintering. Experiments must be made at a higher temperature, using powders with little distortion of the crystal lattice. Meanwhile, it can be stated only that the effect of an additional mechanism is not easy to observe and that the kinetic laws expressed by Eqs. (III-5) and (XI-6) hold true at relatively high temperature even after sintering for a very long time.

*The calculations were made, as described in Chapter III, by means of points corresponding to 0, 0.5, and 2 h of isothermal sintering.

Another question that needs further research is the connection between the densification process and the condition (structure) of the pore surfaces. The absence of any direct effect of surface migration of the substance on the reduction in volume of pores does not imply that the condition of the surface, characterized by roughness and surface defects, has no effect on the densification process. Means of activating sintering (acceleration of densification) by the effect of etchants on the pore surfaces, inducing an increase of submicroscopic roughness, are well known – adding HCl to the furnace atmosphere [143] or dissociation products of NH_4Cl and NH_4F [144], preliminary oxidation of compacts or powders, inducing the formation of easily reduced oxides [145, 146]. The possibility of such treatment points to some relationship between the condition of the surface and the densification rate. Since the effect of the surface condition is more readily associated with processes occurring on the surface, the existence of such a relationship is often taken as an indication of the role of surface migration of the substance in the process of sintering. We have already presented objections to this viewpoint. Let us recall that surface migration can facilitate reduction of the pore surface but not a change in the volume of pores.

There are no signs of the participation of surface migration in isothermal densification, but nevertheless the effect of gaseous etching agents on the surface leads to acceleration of densification. There remains the hypothesis that a change in the condition of the surface has an effect on the rate of bulk flow of the substance. Numerous data have been accumulated indicating that etching agents or surface-active substances affect the surface and the rate of bulk deformation. Above all, this refers to phenomena reflecting the Rebinder effect – the effect of surface-active substances on deformation under loads exceeding the yield strength [147]. An effect of the surface condition and the composition of the gaseous medium was observed also in high-temperature creep [148, 149]. Kuczynski and coworkers found that the rate of formation of necks during sintering of aluminum oxide pellets in hydrogen was higher than in air. They found that regardless of the evident effect of the gaseous medium the kinetics of the process of coalescence is determined by bulk diffusion flow [150, 151]. Unevennesses induced by etching should increase the concentration of vacancies in the surface during heating to high temperature. But it cannot be assumed that under these conditions more complex mobile defects occur in the surface that are responsible for acceleration of bulk flow (let us recall that very large gradients of capillary pressure are created around pits of small radius).

Thus, it seems probable that etching the surface of pores has an effect on the interaction of surface and bulk defects and leads to an increase in the concentration of defects in the sintered body. It must be assumed that many observations on the effect of the medium on the densification process are explained by this interaction.*

In discussing questions of the kinetics of densification we disregarded the presence of grain boundaries. No signs of the effect of grain boundaries on the mechanism of densification associated with the existence of active defects are evident (this statement does not apply to the diffusion mechanism of densification at temperatures close to the melting point, when the grain boundaries serve as vacancy sinks).

The effect of the boundaries may also appear in the fact that bulk flow of the substance can be included as one of the mechanisms of deformation slip along grain boundaries. Slip along boundaries is more probable in the beginning of sintering, when the area of contact between particles of the powder is small and the capillary forces are substantial. However, there are no signs of any kind that this mechanism participates in isothermal sintering. This is indicated in particular by the single kinetics of den-

*In fact, the phenomenon may be more complex. The connection of the effective value or apparent value of surface tension and the roughness of the surface has still not been analyzed. Meanwhile, the possibility is not excluded that there is a change in the value with a rough surface, particularly with "unsymmetrical" roughness (with different curvatures of depressions and protrusions). If the radius of depressions is smaller than the radius of protrusions then a component of the capillary forces may appear that is directed over the whole surface, leading to an apparent increase of surface tension. If such an effect is possible then it affects the densification rate. The circumstance that the process of elimination of defects (and reduction of flow) is described by a simple kinetic equation compels the assumption that the increase in the apparent value of surface tension, if it occurs, is combined with an increase of flow, and both effects are described by very similar mathematical formulations. This may be possible in the presence of a constantly retained relationship between the roughness of the surface and the lattice defects. The assumption of the interaction of surface and bulk defects of the lattice and the "dynamic equilibrium" between them was published earlier [39].

sification at the beginning and end of prolonged sintering, when the probability of slip becomes very small. Finally, even if slip along boundaries takes some part in deformation it should be kept in mind that slip becomes possible only with simultaneous bulk deformation of grains, which in this case will also be a process controlling the course of densification. Therefore slip along boundaries (if it occurs) is not reflected in the kinetic laws of densification.

Generalizations of the phenomenological theory make it possible to digress from the complex character of the processes occurring in sintering. Despite the simplified interpretation of elementary processes, the phenomenological theory gives an explanation of a broad group of phenomena observed in sintering. Along with this, the phenomenological theory explains the reason for the notable differences in the values of the "activation energy of sintering" found in different investigations for one metal or another. The term "activation energy of sintering" has no definite physical meaning because the course of densification is determined by at least two processes with different values of the activation energies. The value of the apparent activation energy of sintering will vary with the relative development of the processes and their interaction under various sintering conditions. Therefore the value of the apparent activation energy will differ, depending on the conditions in which it is determined, even for a single original powder. In determining the activation energy from the change in the densification rate before and after the change from one temperature of isothermal sintering to another the densification rates and the value of the activation energy found from them vary with the rate of increase in temperature. With a slow change from low to high temperature the acceleration of densification observed is smaller than with a rapid rise in temperature because of the substantial reduction in the concentration of defects. Therefore, with a slow rise in temperature the value of the apparent activation energy of sintering is lower, as was observed in experiments with sintering of copper compacts with the temperature increasing at the rate of 3 deg/min between periods of isothermal sintering [152]. The apparent value of the activation energy of sintering for copper was found to be 49,000 cal/g-atom. Similar experiments made by the author gave values of the apparent activation energy from 50,000 to 65,000 cal/g-atom, depending on the rate of increase in temperature.

In determining the activation energy with the time necessary to reach a certain density (or certain level of mechanical or physical properties), quite different values may be obtained for the same powder. The activation energy in this case is determined from the values τ_1, T_1 and τ_2, T_2 by the equation

$$E_{\kappa} = \frac{R \log \frac{\tau_1}{\tau_2}}{0.434 \left(\frac{1}{T_1} - \frac{1}{T_2} \right)}. \quad \text{(XIII-17)}$$

From (XI-6) we can find τ for the given ratio v/v_{in}:

$$\tau = \frac{\left(\frac{v_{in}}{v}\right)^{\frac{a}{b} \exp \frac{\Delta E}{RT}} + 1}{aN_{in} \exp\left(-\frac{E_a}{RT}\right)}.$$

From this equation it follows that the isothermal sintering time required to attain a given value of v/v_{in} varies with the initial concentration of defects and, consequently, with the conditions of heating to isothermal sintering temperature. Also, the volume of pores at the beginning of isothermal sintering at different isothermal sintering temperatures is different (let us recall that a substantial reduction in volume of pores also occurs in the process of heating, the degree of reduction depending on the temperature to which the sintered body is heated). Therefore, the value of the apparent activation energy in this case also depends greatly on the experimental conditions – the rate of heating to constant temperature, the selected temperature range, etc. This circumstance and also the effect of the method of obtaining the powder explain the quite different values of the apparent activation energy of sintering obtained in different investigations – from 35,000 cal/g-atom [132] to 80,000 cal/g-atom [12] for copper, for example.

Depending on the experimental conditions and mainly on the heating rate, the value of the apparent activation energy of sintering may be above or below the activation energy of self-diffusion. Values close to the activation energy of self-diffusion obtained at several average rates of increase in temperature are far from being the case for all powders.

The calculation of the change in viscosity in the process of sintering used by several authors does not give the activation energy a more definite physical meaning. The activation energy of sintering was determined [123, 124, etc.] from the temperature dependence of the calculated value of the viscosity, which was found by extrapolation of the experimental values of viscosity to $\tau = 0$. Meanwhile, the flow rate just at the beginning of sintering varies greatly with the concentration of defects, which is intimately associated with the conditions of the experiment and above all with the heating rate and sintering temperature. In addition, calculating the change in viscosity by means of the relationship $\eta = \eta_0 + \omega\tau$ or $\eta = \eta_0(1 + \xi\tau)$ [123, 124] does not take into account the viscosity of an equiaxed crystal (with $\tau \to \infty$ and $\eta \to \infty$, but not to the final value corresponding to the viscosity of an equiaxed crystal). Consequently, the calculation disregards precisely that value of the viscosity for which the activation energy is compared with the activation energy of self-diffusion.

It is difficult to regard viscosity as a complex compound value and much more convenient to consider it as a concept of flow. Its value can be added up from two components – flow of the diffusion type and flow associated with active lattice defects. At high temperature in an almost equiaxed crystal the first component has a substantial value and the second is close to zero. With a large concentration of defects the second component far exceeds the first and the effect of the diffusion component can be neglected. In the intermediate region (with a not too large concentration of defects and moderately high temperature) it may prove necessary to calculate the effects of both components. In this case it should be kept in mind that the activation energy of different mechanisms of flow may be different.*

As already noted, under sintering conditions ordinarily encountered in practice the flow does not appear to be complex and the flow of the substance at a given temperature is determined mainly by its dependence on the concentration of defects. The temperature dependence of this type of flow can be established fairly precisely only when steps are taken to eliminate (or reduce) the distorting effect of a variable concentration of defects; any other value of the activation energy will be of random character. This

*It is quite possible that further development of the theory of flow of crystalline bodies under small loads will result in substantial corrections. In particular, it may prove that the "high-temperature" and "low-temperature" mechanisms of flow are not so different in principle.

refers also to the coincidence of the activation energy of sintering with the activation energy of self-diffusion found in some work. The chance character of this coincidence is easily demonstrated by a simple experiment – it is usually sufficient to change the rate of heating to a given sintering temperature in order to obtain a quite different value of the activation energy.

The phenomenological theory explains the reason for the different apparent values of the activation energy of sintering: The course of densification (reduction in volume of pores) is determined by the interaction of two phenomenologically elementary processes – the flow of the crystalline substance due to the lattice defects and the change in the concentration of defects with time – with different activation energies. Therefore the apparent activation energy of sintering for a given powder may vary within wide limits, depending on the experimental conditions in which the temperature dependence of shrinkage during sintering was determined.

Phenomenological analysis of elementary processes makes it possible to explain the basic characteristics of densification. Although the physical nature of the elementary processes remains for the most part unknown, it can be stated that exact formulation of the more common laws of densification will facilitate further development of the theory. The validity of the physical interpretation can now be checked by comparing the conclusions of the physical theory with the mathematical description of the kinetics of densification, which is the most general expression of the phenomenological nature of this phenomenon.

Chapter XIV

Phenomenological Generalizations and Sintering Practice

Densification during sintering of metal powders, if its development is not disturbed by other processes or external influences, obeys the simple laws described above. These laws also extend to powders of carbides, nitrides, and borides of refractory metals.

The universality of the kinetic relationship expressed by Eq. (III-5) is characterized by the following data: With sintering under conditions ensuring undisturbed densification (i.e., with an initial density corresponding to a constant value of v_s/v_p at different d_p) the author has never observed a single sample of copper, nickel, iron, cobalt, silver, tungsten carbide, or titanium carbide with a systematic deviation from the curve matching Eq. (III-5).* In addition, the magnitude and probability of random deviations, as follows from Fig. 12, are fairly small.

The strict conformity of the variation of the reduction in volume of pores with kinetic equations (III-5) and (XI-6) makes it possible to use them to calculate the probable densification under given sintering conditions or for preliminary determination of the sintering conditions that would ensure obtaining parts with a given porosity. With mass production of powder the kinetic constants of the powder (with consistent production conditions) vary little and the average values can be used in calculations. The method of determining the constants of a powder and the reduction in volume of pores is given briefly in the Appendix.

*We refer to the isothermal part of sintering.

From the value of v_s/v_p found for specific sintering conditions one can calculate the absolute and relative density of the sintered body or the porosity, expressed in fractions or percent of the volume of the body [3]. To make these calculations it is necessary to know the density or porosity of the green compact.

The density before sintering is determined by

$$d_s = \frac{d_p d_c}{d_p + (d_c - d_p)\frac{v_s}{v_p}}, \tag{XIV-1}$$

where d_p is the density before sintering and d_c is the density of the solid metal, g/cm^3. The relative density is determined by means of a similar equation

$$D_s = \frac{D_p}{D_p + (1 - D_p)\frac{v_s}{v_p}}. \tag{XIV-2}$$

To determine the porosity, expressed in fractions of the volume of the body, one uses

$$P_s = \frac{P_p \cdot \frac{v_s}{v_p}}{1 + P_p\left(1 - \frac{v_s}{v_p}\right)}, \tag{XIV-3}$$

where P_p is the porosity of the green compact.

Finally, the bulk shrinkage, being the ratio of the volumes after and before sintering, is found by

$$\frac{V_s}{V_p} = D_p + (1 - D_p)\frac{v_s}{v_p} \tag{XIV-4}$$

or

$$\frac{V_s}{V_p} = 1 - P_p\left(1 - \frac{v_s}{v_p}\right). \tag{XIV-5}$$

All these equations are easily derived from Eq. (I-1) and ordinary relationships between the density, porosity, and volume porosity of the body.

Since kinetic equation (XI-6) describes densification in the pure form, with considerable development of processes preventing densification the calculation from the constants of the powder becomes inexact or impossible. Therefore it is important to establish the conditions for each powder in which the development of expansion has a substantial effect on the course of densification.

Let us recall that of the two main reasons for expansion of the sintered body – elastic-plastic deformation of particles with removal of strain hardening, due to the elastic-plastic aftereffect [10], and an increase of the gas pressure in closed pores – the latter has the greatest effect on densification. The effect of expansion becomes notable after some critical value of the initial density is exceeded. The method of determining the value of the critical density consists in plotting v_s/v_p vs. d_p from experimental data. The density of the compact d_p, corresponding to the beginning of rapid increase of v_s/v_p with increasing d_p, is the critical value of the density d_{cr} for given sintering conditions (see Chapter II for a more detailed discussion).

The relative value of the critical density $D_{cr} = d_{cr}/d_c$ depends both on the properties of the powder and the sintering conditions. Closed pores are easily formed during compacting of powders of ductile metals. Therefore the critical value of the density for powders of ductile metals (copper, silver, gold, and others) is reached at the lowest compacting pressure. For powders of metals in the iron group the critical value of the density is attained at relatively high pressures, and at still higher pressures for compacting powders of difficult-to-deform metals and alloys (alloy steels, for example). When powders of materials not susceptible to plastic deformation under normal conditions are compacted (carbides of refractory metals, for example) the critical value of the density is not attained even at the very high compacting pressures used in standard production or in laboratory practice (see Fig. 9).

It should be noted that the formation of closed pores does not always lead to disruption of the normal course of densification. With low saturation of the powder with gases, no impurities of reduced oxides, and slow heating ensuring (not always, but in some cases) elimination of a substantial portion of the sorbed gases in the period where interconnected pores are formed, the course of densification conforms with the kinetic laws inherent to it down to small values of porosity.

Thus, the value of v_s/v_p calculated from experimental sintering data for molybdenum powder with different densities given in [153] is quite constant despite the very low residual porosity (2-7%).* With such a small total porosity the fraction of closed pores should be substantial, which, however, does not alter the constant value of v_s/v_p at different d_p. The phenomenon of expansion during sintering of refractory metals (tungsten, molybdenum, tantalum, etc.) is hardly observed, probably because of the high sintering temperature, which ensures fairly complete elimination of sorbed gases before the beginning of isothermal sintering.

For a given powder the critical density changes under different sintering conditions, depending on the temperature (to the least extent) and sintering time. The value of the critical density is affected to some extent by the rate of increase in temperature before the beginning of isothermal sintering. The variation of the critical density with the temperature is shown in Fig. 43 for several powders. The lines limiting the bottom of the hatched regions correspond to rapid increase of temperature (60-100 deg/min) and holding for 2 h. With a smaller rate of increase in temperature and with a shorter holding time the critical density increases within the limits of the hatched zones.

The variation of D_{cr} with T serves as an additional characteristic of the powder and is of considerable practical value. From the graph one can determine directly the value of the initial density and the sintering temperature that ensure undisturbed densification and also the conditions in which the retarding effect of the gas pressure is possible (hatched area) or the process of expansion develops intensively (above the hatched area). Such graphs are useful for commercial grades of powder. If the conditions of manufacturing the part (initial density and sintering temperature) correspond to the field of the graph below the hatched zone then calculations from the kinetic constants of the powder give fairly accurate results. It should be kept in mind only that the plot of D_{cr}

*The value of v_s/v_p was calculated from experimental data obtained by Grube and Schlecht, given in [1].

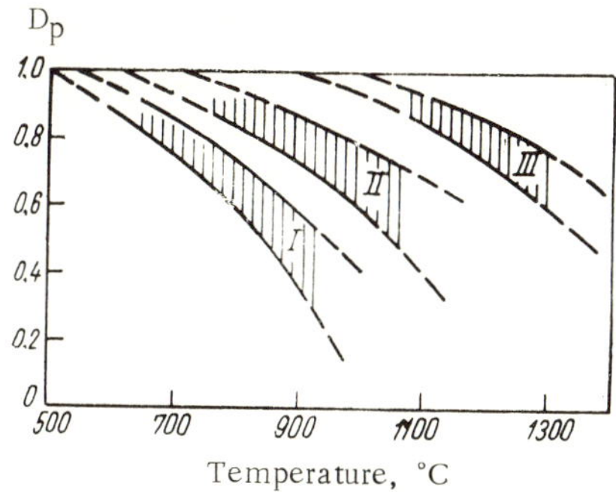

Fig. 43. Variation of relative critical density of compacts with sintering temperature for different powders. I) Active nickel powder obtained by reduction of oxides; II) copper powder obtained by reduction of oxides; III) iron powder obtained by reduction of ground scale. The hatched areas show the possible values of the critical density; the dashed lines indicate extrapolation from experimental data.

vs. T should be constructed for sintering in the same gas in which the part of the given powder will be sintered.

If at a given temperature the value of the relative initial density D_p falls within the hatched area (see Fig. 43) then the development of expansion must be considered a possibility. In this case the calculation of v_s/v_p from the constants of the powder makes it possible to determine the extent of the effect of the gas pressure in closed pores on the change in the volume of pores under given sintering conditions (the more the experimentally determined value of v_s/v_p exceeds the calculated value, the greater the effect of expansion on the final volume of pores). This calculation is important, since measures intended to further increase the density of sintered parts (when the density is insufficient after sintering tests) depend on the fact that the porosity is determined mainly by it – by the inevitable incompleteness of shrinkage in conformity with the kinetic laws or by the opposing force of the gas pressure in closed pores.

The usual measures taken to reduce porosity – increasing the temperature, increasing the sintering time, or using compacts with a higher initial density – are successful only if the densification process is not disrupted by accompanying expansion. With considerable development of expansion a higher sintering temperature or increase of the initial density induces only an increase of the porosity in the sintered body. To change the sintering conditions leading to an increase of porosity it is necessary to reduce the initial density below the level corresponding to the critical density for the given sintering conditions. Reducing the initial density in combination with an increase of sintering temperature makes it possible in some cases to reduce the porosity of the sintered body substantially. As an example let us take sintering of copper powder obtained by reduction of oxides in hydrogen at 500°C. After sintering at 800°C the porosity was 18% for a compact with a relative density $D_p = 0.74$. Increasing the sintering temperature to 950°C not only did not increase the density of the part but led to an increase of porosity to 23%. This could be foreseen, since the relative critical density D_{cr} of this copper powder was equal to 0.68 (for sintering at 800°C). After reduction of the initial porosity D_p to 0.64 (below the critical) the porosity was only 9% after sintering at 950°C.

With considerable development of expansion, substantially interfering with densification, the reduction of the initial density in combination with an increase of the sintering temperature is the simplest means of increasing the density of the part.

The development of expansion also promotes high gas saturation and a tendency to exceedingly rapid densification characteristic of active powders with spongy particles. It is often useful to "coarsen" such powder somewhat (reduce its activity and reduce the specific surface) by means of annealing in hydrogen. Coarsening of the powder in combination with an increase of sintering temperature also makes it possible in some cases to obtain parts that are more dense than parts of the original active powder. This is confirmed by the position of the curves in Fig. 3. The value of d_s for comparable points on the curves for different powders is found from the values of v_s/v_p and d_{in} by Eq. (XIV-1).

The furnace atmosphere used in sintering has a greater or lesser effect on the value of the kinetic constants for different powders (affecting mainly a/b) and a small effect on the value of the critical density of the compacts. Sintering in vacuum usually increases densification somewhat with interconnected pores but only slightly broadens the region of density where v_s/v_p is constant. The critical density is increased substantially by use of compacting in vacuum. Convincing evidence of this can be found in [9]. The closed pores formed in vacuum compacting were not filled with gas. Evacuation also leads to elimination of part of the sorbed gases. Thus, in sintering of compacts pressed in vacuum the gas pressure in the pores is substantially smaller than in the case of ordinary compacts. As was reported in [9], the positive effect of vacuum compacting is retained in sintering both in vacuum and in hydrogen, and the difference in the effect is almost imperceptible. This means that with vacuum compacting one can use ordinary sintering equipment, with hydrogen or another protective atmosphere. Evidently vacuum compacting is an effective means of increasing the density of the part after sintering, since an increase of the critical density makes it possible to increase the initial density and, in conformity with Eq. (XIV-1), the final density. However, it should be pointed out that this refers only to compacts of ductile metals, in which the critical density is attained at commonly used compacting pressures. In compacts of low-ductility metals or materials the critical density is not attained under ordinary compacting conditions and the use of vacuum has no noticeable effect.

The effect of the gas pressure can also be reduced by lowering the heating rate. However, this leads at the same time to reduction of densification during subsequent isothermal sintering, and is often ineffectual for that reason. Only in cases where densification in the absence of the opposing force of the gas pressure is far higher under the given sintering conditions (which can be determined by calculation) can an increase of the heating rate increase the densification.

It is doubtful that the variation of densification during isothermal holding with the rate of the preceding increase in temperature would be of immediate practical use. An increase of the heating rate at the beginning of sintering, making it possible to increase densification somewhat during isothermal holding, may lead to several complications – distortion of the shape ("warping") of parts due to uneven heating, premature formation of closed pores and, as a consequence, reduction of the critical density, etc. As already noted, slower heating to achieve more even densification and to prevent warping of the part may lead to reduction of the total densification.

All the complications associated with the process, of expansion, making it necessary to calculate the critical density of compacts, refer mainly to powders of metals with substantial ductility. Powders of low-ductility metals or carbides are free of these limitations. The course of densification of these powders corresponds to the kinetic laws of pure densification at all compacting pressures encountered in practice.

The graph of the variation of the critical density with temperature is very important for powders of ductile metals and should be used to select the compacting and sintering conditions. The method of determining the critical density was given in Chapter II. Let us note here only that the deviation from constant values of v_s/v_p noted in some cases (a slight increase or decrease of v_s/v_p with increasing d_p) does not interfere at all in determining the value of the critical density, since in this case the change from a linear variation in the region of small densities to a rapid increase of v_s/v_p is clearly visible on the plot of v_s/v_p vs. d_p.

It is sometimes of value in practice to simplify the shape of the pores in the absence of densification. As was noted in Chapter VI, a notable reduction of the surface of pores may continue after the rate of reduction in volume of pores has decreased almost to zero. Since simplification of the shape of pores leads to improvement of the mechanical properties of the sintered body, it can be assumed that a notable change in the mechanical properties may occur during continued sintering after shrinkage has decreased greatly or practically ceased. It must be kept in mind only that the mechanical characteristics may be affected by other processes developing during sintering (grain growth and changes in the substructure), and in some cases the effect of simplifying the shape of the pores may be masked by other effects. In manufacturing filters, prolonged holding at moderately high temperatures may increase the permeability of the filter with hardly any change in the total porosity. In general, the phenomenological laws must be kept in mind in solving practical sintering problems. Calculating the densification characteristics from the constants of the powder is essential for selection of sintering conditions in many cases.

It would be useful to indicate the kinetic constants of a powder for lots produced commercially. Then the consumer would be able, without sintering tests, to determine by calculation the possible densification of the powder under given sintering conditions or to select sintering conditions ensuring parts with a given density. Such calculations can be made with a slide rule in 10-15 min (the calculation of v_s/v_p from the constants of the powder is described briefly in the Appendix).

Phenomenological analysis can also be used to determine more precisely the kinetic laws in the concluding stage of liquid-phase sintering, in the process of which the solid particles coalesce with formation of a skeleton of the high-melting component of the alloys, if a notable fraction of densification falls in this stage of sintering. The application of phenomenological analysis to liquid-phase sintering deserves special consideration. Let us say here only that the additions forming the liquid phase (both soluble and not soluble in the high-melting component of the mixture) do not change the form of the kinetic relationship and affect only certain constants. The addition of a low-melting component, accelerating densification, leads to reduction of v_0 and a lower value of a/b. The reduction of v_0 is reflected in the fact that with participation of liquid phase in sintering a substantial fraction of densification occurs at the beginning of sintering in the period of shifting and redistribution of free particles (the stage of "liquid flow," according to Kingery [154]). For this reason, the reduction in the volume of pores increases before the beginning of even densification due to the coalescence of solid particles. The reduction of a/b reflects the increase of b due to an increase of the effect of capillary force on the densification of the skeleton.

The possibilities of applying phenomenological analysis to single-phase sintering are far from exhausted. It is very probable that the connection of several constants with structural characteristics of the particles of powder will be precisely determined in the future. This refers especially to the values of v_0 and a/b. It can be assumed that the connection between the value of b and the dispersity of the powder, which is clear from phenomenological analysis and confirmed by experiment, will be determined quantitatively.

Phenomenological generalizations cannot be disregarded in formulating a physical theory of sintering. The principal result of phenomenological analysis in its theoretical aspects is as follows: Densification is determined by the development and interaction of two elementary processes with different kinetic characteristics and different activation energies. This explains the pronounced variation of the densification rate with temperature. Therefore the hypotheses advanced to explain densification that are based on the assumption of a single atomic mechanism ensuring both deformation of the crystal and reduction of the deformation rate with time must be rejected as not corresponding to the basic phenomenological characteristic of densification in sintering of metal powders.

There are several problems resulting from phenomenological analysis that must be solved by physical theory. One of them is to determine more exactly the effect on the densification rate of the changing geometry of pores during sintering. The empirical relationship $dv/d\tau = -\xi v$ is completely suitable for phenomenological analysis (it makes it possible to describe both the variation of densification with time at a constant concentration of defects and the relative reduction in volume of pores in compacts with different initial porosities), but it still lacks a strict physical explanation. The use of the theory of viscous flow of an ideally viscous body as a model of a porous body with cylindrical interconnected pores [88] closely approximates the description of the time dependence of densification with constant viscosity, but the deductions from the mathematical relationship obtained do not match a constant relative reduction in volume of pores with different initial densities.

Skorokhod recently proposed as a model of a porous body a statistical mixture of sections of the substance and voids and, using the rheological approach to calculate the macroscopic viscosity of a porous body, found that according to the equation he obtained the rate of reduction in porosity should be proportional to the value of the porosity and some factor depending on the size of the particles and the viscosity of the substance [7]. From the expression obtained it follows directly that P_s/P_p is independent of the initial porosity or density. This relationship is close to a constant relative reduction in volume of pores, but not equivalent to it. The difference between these relationships (P_s/P_p = const and v_s/v_p = const) is considerable with a substantial change in the initial density of the compacts (see Chapter I). On the whole, one can say that the theory approaches a solution to the problem of describing the effect of the geometrical factor. Further refinement of the theory will probably lead only to selection of a model closest to the "structure of interconnected channels" established during isothermal sintering.

Far more complex is the problem of determining the physical nature of the laws connecting flow with the concentration of defects and the elimination of defects. At the present time there is no assurance that a physical treatment of these phenomena is possible on the basis of existing data on the nature of defects, their interactions, and their effect on flow. Further perfection of the sintering theory will depend greatly on the development of a theory of the defective condition of the crystalline substance and, particularly, the physical nature of the defects responsible for acceleration of the flow of a crystalline substance under small loads. Metal physics must explain the remarkable fact that there are common laws governing the change in flow of imperfect crystals with time for crystalline substances quite different in their physical nature – from copper and silver to tungsten and titanium carbides.

It can be expected that further research on the phenomenology of sintering will not only lead to perfection of the phenomenological theory but also make some contribution to the data needed for a more precise and more fully developed theory of flow of defective crystals under small loads. This is an important but still inadequately developed branch of solid-state physics.

Appendix

Method of Determining the Kinetic Constants of the Powder and Calculating the Reduction in Volume of Pores and Other Densification Characteristics from the Constants of the Powder

The course of densification (reduction in volume of pores) during sintering is determined by the constants:

E_a, the activation energy of the elimination of defects, cal/g-atom;

E_b, the activation energy of flow due to the presence of defects, cal/g-atom;

a/b, the ratio of the preexponential factors of the kinetic equations of elementary processes, a dimensionless unit;

aN_0, the kinetic characteristic of the relative concentration of defects in the original powder, h^{-1};

v_0, the relative volume of pores in the sintered body, from which even densification begins that is described by the kinetic equations;

K, a coefficient compensating the systematic error introduced by simplifications in the rough calculations.

I. DETERMINING THE KINETIC CONSTANTS OF THE POWDER

1. Original Data

a) To determine ΔE, a/b, v_0, and K one needs experimental data from several isothermal sinterings for 2 h at different temperatures. The experiments are made so far as possible with the same rate of increase in temperature to isothermal holding temperature.

To calculate the constants we use the value of the ratio of the volume of pores after and before sintering v_s/v_p with isothermal holding for 0, 0.5, and 2 h at two sintering temperatures (for rough calculations) or for several temperatures (for more accurate calculations).

The value of v_s/v_p is found from the experimentally determined values of density before and after sintering d_p and d_s and the density of the solid metal d_c in conformity with the expression

$$\frac{v_s}{v_p} = \frac{d_p(d_c - d_s)}{d_s(d_c - d_p)},$$

or by the data from dilatometric measurements, as described in Chapter III. The values of v_s/v_p for sintering times of 0, 0.5, and 2 h will be designated v_{in}, v_1, and v_2.

(b) To determine the average rate of increase in temperature up to the beginning of isothermal holding α and the time taken as the arbitrary beginning of isothermal sintering τ_0 an idealized temperature graph is plotted from the real variation of T with τ, as described in Chapter XII [see the discussion following Eq. (XII-21) and Fig. 40].

The value of α, deg/h, is found from the slope of the line characterizing the average rate of increase in temperature. The value of τ_0 (h) is equal to the point on the x axis equivalent to the point of intersection of the sloping and horizontal lines on the idealized temperature graph.

If the value of α does not differ more than $\pm 20\%$ from the average value then some average value of α is used in calculating the constant for all isothermal sinterings.

(c) To determine E_a and E_b we need experimental data characterizing the change in volume of pores with a stepped rise in temperature, i.e., the values of v_s/v_p for two periods of isothermal sintering with a rapid change of temperature between them (see Chapter XI, Fig. 30). The length of each sintering period is 2 h. These data can be obtained by sintering in a furnace equipped with a dilatometric apparatus. The change in the volume of pores is calculated from the dilatometric curves [see the discussion preceding Eq. (III-1) in Chapter III].

For the calculation we use the values determined by means of the experiment with a stepped rise in temperature: $v_{in(1)}$, $v_{in(2)}$, the relative volumes of pores for the beginning of the first and second periods; $v_{fi(1)}$ and $v_{fi(2)}$, for the end of the same periods of isothermal sintering; T_1 and T_2, the temperatures of each sintering period (let us recall that v_{in} and v_{fi} are the relative volumes of pores expressed by the ratio v_s/v_p for the corresponding sintering times).

2. Determining the Constants q and m in Eq. (III-5) for Isothermal Curves of $v_s/v_p = f(\tau)$

In the calculation we use the values of v_{in}, v_1, and v_2 for isothermal sinterings found by the method described in Section 1a.

The value of m is found by selecting a value satisfying the equality

$$\frac{\left(\frac{v_{in}}{v_2}\right)^m - 1}{\left(\frac{v_{in}}{v_1}\right)^m - 1} = 4.$$

The selection can be replaced with graphic solution of this equation. The variation of $\varphi(m)$ with m is plotted, where $\varphi(m)$ is the left side of the equality presented above. The point of intersection of the curve plotted from several arbitrarily chosen values of m with the horizontal line drawn for $\varphi(m) = 4$ gives the desired value of m.

Then we determine q:

$$q = \frac{\left(\frac{v_{in}}{v_1}\right)^m - 1}{0.5m}\ \mathrm{h}^{-1} \quad \text{or} \quad q = \frac{\left(\frac{v_{in}}{v_2}\right)^m - 1}{2m}\ \mathrm{h}^{-1}.$$

It is expedient to use both formulas in order to demonstrate the identical value of q obtained both for v_1 and $\tau_1 = 0.5$ h and v_2 and $\tau_2 = 2$ h. This assures us that the calculated values of m are correct.

The values of q and m are found for all isothermal sinterings. In addition, the values of m are found for each of the two periods of isothermal sintering in the experiment with a stepped rise in temperature.

3. Determining ΔE and *a/b*

The variation of log m with 1/T is plotted from the values of m determined as described in Section 2. Lines are drawn along the experimental points, along which one arbitrarily selects two points with coordinates log m_1, $1/T_1$ and log m_2, $1/T_2$:

$$\Delta E = \frac{\log m_1 - \log m_2}{\frac{1}{T_1} - \frac{1}{T_2}} \cdot 4.58 \text{ cal/g-atom}$$

$$\frac{a}{b} = m_1 \cdot \exp\left(-\frac{\Delta E}{RT_1}\right)$$

or

$$\log \frac{a}{b} = \log m_1 - \frac{\Delta E}{4.58 T_1}.$$

The calculation can be checked by determining *a/b* from the same formula but with the values of m_2 and T_2.

For very rough calculation of ΔE and *a/b* one can use two values of m determined for two isothermal sinterings at different temperatures.

4. Determining E_a and E_b

From the experiment with a stepped increase in temperature (Section 2) one uses the values $v_{in(1)}$, $v_{in(2)}$, $v_{fi(1)}$, $v_{fi(2)}$, τ_1, τ_2, T_1, and T_2, which are substituted (in the equation given below) for the values of m_1 and m_2 calculated for the first and second periods of isothermal sintering, as described in Section 2.

First let us find the value of M:

$$M = \frac{\left[\left(\frac{v_{in(2)}}{v_{fi(2)}}\right)^{m_2} - 1\right] \tau_1}{\left[1 - \left(\frac{v_{fi(1)}}{v_{in(1)}}\right)^{m_1}\right] \tau_2}.$$

Then we determine E_a and E_b:

$$E_a = 4.58 \cdot \frac{\log M}{\frac{1}{T_1} - \frac{1}{T_2}} \quad \text{cal/g-atom}$$

$$E_b = E_a + \Delta E \quad \text{cal/g-atom}$$

5. Determining the Values of aN_0, v_0, and K

For all isothermal sinterings one determines the value of aN'_{in} for $v_0 = 1$:

$$aN'_{in} = \frac{qm}{v^m_{in}} \cdot \exp\frac{E_a}{RT} \ \mathrm{h}^{-1}.$$

The values of v_{in}, q, and m were determined in Sections 1 and 2 for each isothermal sintering.

From the values of E_a and T, by means of Table 27 and the interpolation formula (XII-2), one finds the value of function F_T for all sinterings. Then, from the values found for aN'_{in} and F_T one plots the variation of $1/aN'_{in}$ with F_T. Straight lines are drawn through the experimental points. For further calculation one uses the coordinates of two arbitrarily chosen points on the line. From the coordinates of these points one finds the values of $aN'_{in(1)}$, $F_{T(1)}$, $aN'_{in(2)}$, and $F_{T(2)}$.

The position of the line in regard to the origin of coordinates determines the further course of densification.

(a) On the plot of $1/aN'_{in}$ vs. F_T the line passes below and to the right of the origin of coordinates. The value of v_0 is determined (more precisely, its limit value):

$$\log v_0 = \frac{\log aN'_{in(1)} - \log aN'_{in(2)} + \log F_{T(1)} - \log F_{T(2)}}{m_1 - m_2}.$$

where $aN'_{in(1)}$, $F_{T(1)}$ and $aN'_{in(2)}$, $F_{T(2)}$ are values corresponding to the coordinates of the points selected on the line. The values of m_1 and m_2 must be calculated from the values of a/b and ΔE for temperatures corresponding to the values chosen for $F_{T(1)}$ and $F_{T(2)}$ [$m = (a/b)\exp(\Delta E/RT)$, where $\Delta E = E_b - E_a$]. The values of $F_{T(1)}$ and $F_{T(2)}$ should be selected so as to match rounded values of T, in multiples of 50°C.

Then we find K:

$$K = aN'_{\text{in}(1)} \cdot v_0^{m_1} \frac{F_{T(1)}}{\alpha},$$

where α is the average rate of increase in temperature for the initial periods of all isothermal sinterings, deg/h. It is expedient to check the calculation by substituting $aN'_{\text{in}(2)}$, $F_{T(2)}$, and m_2 into the same expression.

If the values of aN'_{in} calculated from the variation of $1/aN'_{\text{in}}$ with F_T differ by more than four orders then the values of aN'_{in} are calculated again for $v'_0 < 1$. In the calculation one uses an arbitrarily chosen intermediate value of v'_0 (0.8 or 0.6, for example). The choice can be considered fortunate if the maximum difference in the values of aN'_{in} does not exceed 10^4 times. In this case the value of aN'_{in} can be calculated from the formula

$$aN'_{\text{in}} = qm \left(\frac{v'_0}{v_{\text{in}}} \right)^{m_1} \cdot \exp \frac{E_a}{RT} h^{-1}.$$

In other respects the calculation is like the preceding: plot $1/aN'_{\text{in}}$ vs. F_T, select points on the line, from the coordinates of which one finds the values of $aN'_{\text{in}(1)}$, $F_{T(1)}$ and $aN'_{\text{in}(2)}$, $F_{T(2)}$ that are used to calculate v''_0 by means of the formula given above for $\log v_0$. The final value of v_0 is found as the product of v'_0 and v''_0:

$$v_0 = v'_0 v''_0.$$

Here, v''_0 was calculated by the formula, while v'_0 was selected as an intermediate value in the calculation of aN'_{in}.

K is found by the formula

$$K = aN'_{\text{in}(1)} \left(\frac{v_0}{v'_0} \right)^{m_1} \frac{F_{T(1)}}{\alpha}.$$

Powders for which this method is used are arbitrarily called "active."

(b) On the plot of $1/aN'_{\text{in}}$ vs. F_T the line passes above and to the left of the origin of coordinates. In this case v_0 is taken as equal to unity. From the values of $aN'_{\text{in}(1)}$ and $aN'_{\text{in}(2)}$ established from the points selected on the plot of $1/aN'_{\text{in}}$ vs. F_T and the corresponding values $F_{T(1)}$ and $F_{T(2)}$ one determines the value of K:

$$K = \frac{F_{T(1)} - F_{T(2)}}{\alpha\left(\frac{1}{aN'_{(2)}} - \frac{1}{aN'_{in(1)}}\right)},$$

where α is the average rate of rise in temperature, deg/h, in the initial periods of all isothermal sinterings.

Then we find aN_0:

$$aN_0 = \frac{aN'_{in(1)}}{1 - \frac{aN'_{in(1)}}{\alpha K} \cdot F_T} \ \mathrm{h}^{-1}.$$

It is expedient to check the calculation by repeating it, using values $aN'_{in(2)}$ and $F_{T(2)}$.

Powders for which this method is used are arbitrarily called "low-activity" powders.

If the object of the calculation is to calculate the curve of Eq. (XII-4) for experimentally established $v_s/v_p = f(\tau)$ in the form of a table or graph then the coordinate of the time of the beginning of the curve for this equation τ_0, i.e., the time arbitrarily selected as the beginning of isothermal sintering, is found by the equation

$$\tau_0 = \tau_{in} - \frac{1 - \left(\frac{v_{in}}{v_0}\right)^m}{qm} \ \mathrm{h}$$

Here, v_{in} and τ_{in} are coordinates of the points taken as the beginning of isothermal sintering in calculating constants q and m. The latter are calculated from experimental data, as described in Section 2. The value of v_0 (volume of pores from which even densification begins) must be previously established for the given powder ($v_0 < 1$ for active powders, $v_0 = 1$ for low-activity powders).

Let us recall that Eq. (XII-4) describes all densification as an isothermal process and that in the section between τ_0 and τ_{in} the actual densification and the calculated densification do not coincide (see Fig. 38). The isothermal sintering time in calculating the value of v_s/v_p by Eq. (XII-4) is found as the difference between the total sintering time, including the initial heating period and holding at constant temperature, and the value of τ_0:

$$\tau = \tau_{tot} - \tau_0.$$

In calculating $v_s/v_p = f(\tau)$ from the constants of the powder the value of τ_0 is determined from the idealized temperature graph, as described in Section 1b.

Note: In those cases where some of the experimental points on the plot of log m vs. 1/T (Section 3) lie far from the interpolated line it is expedient for calculating aN'_{in} (Section 5) to use not the value of m obtained directly from the experimental curve but the value of m calculated from the previously found constants a/b, E_a and E_b for the given isothermal sintering temperature:

$$m = \frac{a}{b} \exp \frac{E_b - E_a}{RT}.$$

II. CALCULATING THE VALUE OF v_s/v_p FROM THE CONSTANTS OF THE POWDER

The original data are the constant of the powder, E_a, E_b, a/b, K, and v_0 (for active powders) or E_a, E_b, a/b, K, and aN_0 (for low-activity powders), the rate of rise in temperature up to the beginning of isothermal sintering α (deg/h), the temperature T (°K), and the total sintering time including the rise in temperature and isothermal holding, τ_{tot} (h).

1. Determining the Value of aN'_{in}

For active powders

$$aN'_{in} = \frac{\alpha K}{F_T} \text{ h}^{-1}.$$

For low-activity powders

$$aN'_{in} = \frac{aN_0}{1 + \frac{aN_0}{\alpha K} \cdot F_T} \text{ h}^{-1}.$$

F_T is found by means of Table 27 and the interpolation formula (XII-2) (see the discussion following this formula and Table 27).

2. Determining the Value of $q_0 m$

$$q_0 m = aN'_{in} \exp\left(-\frac{E_a}{RT}\right) h^{-1}.$$

or

$$\log q_0 m = \log aN'_{in} - \frac{E_a}{4.58T}.$$

3. Determining the Value of m

$$m = \frac{a}{b} \exp \frac{E_b - E_a}{RT}.$$

or

$$\log m = \log \frac{a}{b} + \frac{E_b - E_a}{4.58T}.$$

4. Calculating the Value of v

$$v = v_0 (q_0 m\tau + 1)^{-\frac{1}{m}}.$$

or

$$\log v = \log v_0 - \frac{1}{m} \log (q_0 m\tau + 1).$$

For low-activity powders $v_0 = 1$. For active powders v_0 is a constant of the powder. The isothermal sintering time τ is found as the difference between τ_{tot} and τ_0:

$$\tau = \tau_{tot} - \tau_0 \text{ h}$$

The value of τ_0 is found either directly from the idealized temperature graph as the coordinate of the point of intersection of the sloping and horizontal lines or calculated from the isothermal sintering temperature T, the temperature of the chamber T_{ch}, and the average rate of increase in temperature α:

$$\tau_0 = \frac{T - T_{ch}}{\alpha}.$$

Let us recall once again that the value of v calculated by the kinetic equation characterizes the relative volume of pores in the sintered body expressed in fractions of the original volume of pores in the compact: $v = v_s/v_p$.

III. DETERMINING THE DENSITY, POROSITY, AND SHRINKAGE AFTER SINTERING FROM THE VALUES OF v_s/v_p OBTAINED BY CALCULATION

Along with the value of v_s/v_p, as original data we also use the density of the compact d_p (g/cm^3), the density of the solid metal d_c (g/cm^3), or the porosity of the compact P_p, expressed by the ratio of the volume of pores to the volume of the body, $P_p = v_p/V_p$.

1. Calculating the Density of the Sintered Part

$$d_s = \frac{d_p d_c}{\frac{v_s}{v_p}(d_c - d_p) + d_p} \text{ g/cm}^3$$

or

$$d_s = \frac{(1 - P_p)\, d_c}{1 - P_p\left(1 - \frac{v_s}{v_p}\right)} \text{ g/cm}^3$$

2. Calculating the Porosity of the Sintered Part

$$P_s = \frac{\frac{v_s}{v_p}(d_c - d_p)}{\frac{v_s}{v_p}(d_c - d_p) + d_p}.$$

or

$$P_s = \frac{P_p \cdot \frac{v_s}{v_p}}{1 - P_p\left(1 - \frac{v_s}{v_p}\right)}.$$

3. Calculating Shrinkage during Sintering

The bulk shrinkage V_s/V_p, i.e., the ratio of the volumes of the body after and before sintering, is found by the equation

$$\frac{V_s}{V_p} = \frac{v_s}{v_p}\left(1 - \frac{d_p}{d_c}\right) + \frac{d_p}{d_c}$$

or

$$\frac{V_s}{V_p} = 1 - P_p\left(1 - \frac{v_s}{v_p}\right).$$

The linear shrinkage l_s/l_p, expressed by the ratio of the linear dimensions after sintering to the dimension before sintering, is equal to

$$\frac{l_s}{l_p} = \left(\frac{V_s}{V_p}\right)^{\frac{1}{3}}.$$

It should be kept in mind that this means is used to calculate only the average linear shrinkage. The actual shrinkage may differ somewhat in different directions.

Literature Cited

1. V. A. Ivensen, Zh. Tekh. Fiz., 17(11):1301 (1947).
2. M. Yu. Bal'shin, Powder Metallurgy [in Russian], Metallurgizdat (1948).
3. V. A. Ivensen, Zh. Tekh. Fiz., 17(11):1305 (1947).
4. V. V. Skorokhod, Poroshkovaya Met., No. 2, p. 14 (1961).
5. J. K. Mackenzie and R. Shuttleworth, Proc. Roy. Soc., 62:833 (1949).
6. L. D. Landau and E. M. Lifshits, Mechanics of Continua [in Russian], Gostekhizdat (1954).
7. V. V. Skorokhod, Poroshkovaya Met., No. 12, p. 19 (1968).
8. V. A. Ivensen, Zh. Tekh. Fiz., 18:1290 (1948).
9. B. Ya. Pines and A. F. Sirenko, Izv. Vuzov., Ser. Fiz., 1:23 (1960).
10. Ya. B. Fridman, Mechanical Properties of Metals [in Russian], Oborongiz (1952).
11. G. F. Huttig, Z. Elektrochem., 54(1):89 (1950).
12. P. Duwez, Atom Movements, ASM, New York (1951), p. 192.
13. V. A. Ivensen, Zh. Tekh. Fiz., 20(12):1490 (1950).
14. Ya. I. Frenkel', Zh. Éksperim. Tekh. Fiz., 16:29 (1946).
15. A. J. Shaler and J. Wulff, Phys. Rev., 73(8):926 (1948).
16. R. A. Andrievskii and I. M. Fedorchenko, Problems of Powder Metallurgy and Strength of Metals [in Russian], Izd. AN UkrSSR, Kiev (1958), No. 6, p. 19.
17. S. Mäkipirtti, Acta Polytechnica Scand., 265(10):1 (1959).
18. M. Tikkanen, Planseeber. Pulvermetallurgie, 11(2):(1963).
19. N. C. Kothari, Powder Metallurgy, 7(14):251 (1964).
20. M. Yu. Bal'shin, in: Powder Metallurgy [in Russian], Izd. AN SSSR, Kiev (1954), p. 10.

21. M. Yu. Bal'shin, Zh. Tekh. Fiz., 22:685 (1952).
22. R. L. Coble, J. Amer. Ceram. Soc., 41:55 (1958).
23. G. Cizeron and P. Lacombe, Rev. Met., 52:771 (1955).
24. W. D. Kingery and M. Berg, J. Appl. Phys., 26:1205 (1955).
25. A. Okamura, I. Masuda, and S. Kikuta, Sci. Rep. RJTU, Japan, A1:1 (1949).
26. O. Takasaki, Sci. Rep. RJTU, Japan, A5:1 (1953).
27. G. Bockstiegel, J. Metals, 8(5)(Sect. 2):580 (1956).
28. G. A. Gich, Usp. Fiz. Met., 1:120 (1956).
29. D. L. Johnson and I. B. Cutler, J. Amer. Ceram. Soc., 46(11): 541 (1963).
30. T. Vasilos and J. T. Smith, J. Appl. Phys., 35:215 (1964).
31. R. L. Coble, J. Appl. Phys., 32:787 (1961).
32. B. Ya. Pines, Usp. Fiz. Nauk, 52:501 (1954).
33. B. Ya. Pines, Zh. Tekh. Fiz., 23:2077 (1953).
34. S. Scholz, Planseeber. Pulvermetallurgie, Vol. 11 (1963).
35. R. D. Carlsson and F. F. Westermann, Planseeber. Pulvermetallurgie, 10(1):15 (1962).
36. J. D. McClelland and D. L. Whitney, Planseeber. Pulvermetallurgie, 10:131 (1962).
37. G. V. Samsonov and M. S. Koval'chenko, Poroshkovaya Met., No. 1, p. 20 (1961).
38. M. Yu. Bal'shin and A. A. Trofimova, Poroshkovaya Met., No. 1, p. 82 (1961).
39. V. A. Ivensen, Zh. Tekh. Fiz., 20:1483 (1950).
40. I. M. Fedorchenko and R. A. Andrievskii, Fundamentals of Powder Metallurgy, Izd. AN UkrSSR, Kiev (1961), p. 238.
41. Ya. E. Geguzin, Fiz. Metal. Metalloved., 6:650 (1958).
42. G. C. Kuczynski, J. Appl. Phys., 20:1160 (1949).
43. Ya. E. Geguzin, Dokl. Akad. Nauk SSSR, 92(1):45 (1953).
44. G. C. Kuczynski, J. Metals, 185:169 (1949).
45. I. Zaplatynskyj, Planseeber. Pulvermetallurgie, 6(3):89 (1958).
46. A. L. Pranatis and L. Seigle, Powder Metallurgy, Inter. Pub., New York (1961), p. 31.
47. B. Ya. Pines, Zh. Tekh. Fiz., 16(6):737 (1946).
48. B. Ya. Pines, Trudy Khar'kov. Gosunivers., Fiz. Otdel., 3:65 (1955).
49. G. C. Kuczynski, Acta Met., 4:58 (1956).
50. H. F. Fischmeister and R. Lahn, 11 Internat. Pulvermet. Tagung, Eisenach, 1961, Abhand. Deutsch. Akad. der Wissensch., Berlin (1962), p. 93.

51. N. Cabrera, J. Metals, 188:667 (1950).
52. F. Thümmler, Fortschritte der Pulvermetallurgie, F. Eisenkolb and F. Thümmler, eds., Akademie Verlag, Berlin, Vol. 1 (1963), p. 329.
53. D. L. Johnson and I. B. Cutler, J. Amer. Ceram. Soc., 46(11): 541 (1963).
54. I. M. Fedorchenko and R. A. Andrievskii, Poroshkovaya Met., No. 1, p. 9 (1961).
55. V. A. Ivensen, Zh. Tekh. Fiz., 22:677 (1952).
56. V. A. Ivensen, in: Problems of Powder Metallurgy [in Russian], Izd. AN UkrSSR, Kiev (1955), p. 130.
57. I. M. Fedorchenko and N. V. Kostyrko, Fiz. Metal. Metalloved., 10(1):75 (1960).
58. R. I. Garber and S. S. D'yachenko, Zh. Tekh. Fiz., 22:1097 (1952).
59. S. S. D'yachenko, Trudy Khar'kov. Politekhn. Inst., Ser. Inzhen.-Fiz., 14:179 (1958).
60. W. F. Witzig, J. Appl. Phys., 23:1263 (1962).
61. F. E. Faris, Internat. Conf. on Peaceful Uses of Atomic Energy, Geneva, 1955, IAEA, Vienna (1955), paper No. 747.
62. E. R. Jones, W. Munro, and N. H. Hancock, J. Nucl. Energy, 1:76 (1954).
63. Alfred J. Kennedy, Processes of Creep and Fatigue in Metals, John Wiley, New York (1965).
64. H. L. Burghoff, Proc. Amer. Soc. for Test. Mater., 42:668 (1942).
65. H. Udin, A. J. Shaler, and J. Wulff, J. Metals, 1:186 (1949).
66. F. H. Buttner, E. R. Funk, and H. Udin, J. Metals, 4:401 (1952).
67. P. Clark and J. White, Trans. Brit. Ceram. Soc., 49:305 (1950).
68. A. Williams, Symposium on Powder Metallurgy, London, 1954, Spec. Rep. 58, The Iron and Steel Inst. (1956), p. 112.
69. G. A. Meerson, in: Problems of Powder Metallurgy [in Russian], Izd. AN UkrSSR, Kiev (1955), p. 16.
70. C. Herring, in: The Physics of Powder Metallurgy, McGraw-Hill, New York (1951), p. 143.
71. C. Herring, J. Appl. Phys., 21:437 (1950).
72. M. S. Koval'chenko and G. V. Samsonov, Poroshkovaya Met., No. 2, p. 3 (1961).
73. V. A. Ivensen, Fiz. Metal. Metalloved., 6:370 (1958).

74. Ya. E. Geguzin, Physics of Sintering [in Russian], Nauka (1967)
75. Ya. E. Geguzin, Fiz. Metal. Metalloved., 9(6):842 (1960).
76. A. D. Le Kler, Usp. Fiz. Met., 1:224 (1956).
77. F. N. Nabarro, Repts. Conf. on Strength of Solids, Phys. Soc., London (1948), 75.
78. D. McLean, in: Vacancies and Point Defects [Russian translation], V. M. Rozenberg, ed., Metallurgizdat (1961), p. 197.
79. H. Wilsdorf and D. Kuhlmann, Z. Angew. Phys., 4:361 (1952).
80. H. Wilsdorf and D. Kuhlmann, Acta Met., 1:394 (1953).
81. F. W. Young, J. Appl. Phys., 32(192):1815 (1961).
82. N. F. Mott and F. Nabarro, Repts. Conf. on Strength of Solids, Phys. Soc., London (1948), p. 1.
83. A. H. Cottrell, Dislocations and Plastic Flow in Crystals, Clarendon Press, Oxford (1958).
84. C. L. Smith, Proc. Phys. Soc., London 61:201 (1948).
85. K. K. Ziling, Fiz. Tverd. Tela, 7:881 (1965).
86. J. Weertman, J. Appl. Phys., 26:1213 (1955).
87. N. T. Mott, Symposium on Creep and Fracture of Metals at High Temperatures, London, 1954, Stat. Office, London (1956), p. 21.
88. Ya. E. Geguzin, Dokl. Akad. Nauk SSSR, 92:45 (1953).
89. W. D. Kingery, Introduction to Ceramics, John Wiley, New York (1964).
90. J.C. Bokros and A. S. Schwartz, Trans. AIME, 239:7 (1967).
91. W. Kauzmann, Metals Technology, No. 6, p. 317 (1941).
92. M. Cook and T. L. Richards, J. Inst. Metals, 73:1 (1946).
93. W. G. Burgers, Repts. Conf. on Strength of Solids, Phys. Soc., London (1948), p. 134.
94. D. Kuhlmann, Z. Phys., 124:468 (1947).
95. D. Kuhlmann, Proc. Phys. Soc., A64:140 (1951).
96. D. Kuhlmann, Z. Metallk., 40:241 (1949).
97. R. Becker, Phys. Z., 24:919 (1925).
98. A. H. Cottrell and V. Aytekin, J. Inst. Metals, 77:389 (1950).
99. W. Gerlach and W. Hartuagel, Sitzungsbericht Bayer. Akad. Wissenschaft (1939), p. 97.
100. G. Masing and J. Raffensieper, Z. Metallk., 41:65 (1950).
101. Creep and Recovery, ASM, Cleveland (1957).
102. R. Drouard, Trans. AIME, 197:1226 (1953).
103. V. Vand, Proc. Phys. Soc., 55:223 (1943).
104. W. E. Parkins, G. J. Dienes, and F. W. Brown, J. Appl. Phys., 22:1012 (1952).

105. W. Primak, Phys. Rev., 100:1677 (1955).
106. J. R. McDonald, J. Chem. Phys., 36:345 (1962).
107. A. S. Nowich and B. S. Berry, Acta Met., 10:312 (1962).
108. G. W. Brindley, Repts. Conf. on Strength of Solids, Phys. Soc., London (1948), p. 95.
109. D. Bowen, R. R. Eggleston, and R. H. Kropschot, J. Appl. Phys., 23:630 (1952).
110. B. F. Decker and D. Harker, Trans. AIME, 188:181 (1950).
111. E. C. Perryman, Acta Met., 2:26 (1954).
112. E. C. Perryman, J. Metals, 7:1053 (1955).
113. M. A. Yakovlev, Inzhen. Fiz. Zh., No. 11, p. 2 (1959).
114. V. I. Il'ina et al., in: Problems of Metal Science and Physics of Metals, Proceedings of the Central Scientific-Research Institute of Ferrous Metallurgy [in Russian], Metallurgizdat (1952), p. 3.
115. A. Van den Benkel, Acta Met., 11:93 (1963).
116. K. A. Osipov, Some Activated Processes in Metals [in Russian], Izd. AN SSSR (1962).
117. A. W. Overhauser, Phys. Rev., 90:393 (1953).
118. C. J. Meechan, Phys. Rev., 103:1193 (1956).
119. R. Suhrman and H. Schnackenberg, Z. Elektrochem., 47(3): 277 (1941).
120. J. W. Kauffmann and J. S. Kockler, Phys. Rev., 97:555 (1955).
121. J. E. Baurle, Phys. Rev., 102:1182 (1956).
122. A. Damask and G. J. Dienes, Point Defects in Metals, Gordon (1964).
123. V. V. Skorokhod, Poroshkovaya Met., No. 3, p. 3 (1961).
124. M. S. Koval'chenko and G. V. Samsonov, Poroshkovaya Met., No. 5, p. 7 (1961).
125. L. Ramquist, Powder Metallurgy, 9:26 (1966).
126. D. E. Barke and D. Tariball, Usp. Fiz. Met., 1:365 (1956).
127. J. E. Burke, Trans. AIME, 180:73 (1949).
128. P. E. Evans and D. W. Ashall, Intern. Powder Metallurgy, 1:32 (1965).
129. J. R. McEwan, J. Amer. Ceram. Soc., 45:37 (1962).
130. H. Hausner and R. King, Planseeber. Pulvermetallurgie, 8: 28 (1960).
131. M. Yu. Bal'shin, Zh. Tekh. Fiz., 22:678 (1952).
132. I. M. Fedorchenko, Izv. Akad. Nauk SSSR, OTN, No. 3, p. 393 (1953).
133. V. V. Skorokhod, Proceedings of the VII All-Union Conference

on Powder Metallurgy [in Russian], Armyanskii Komitet Poroshkovoi Metallurgii, Erevan (1964), p. 95.
134. R. A. Andrievskii and I. M. Fedorchenko, Inzhen. Fiz. Zh., 3:83 (1960).
135. G. J. Dienes, Phys. Rev., 91:1283 (1953).
136. J. Nihoul, Phys. Status Solidi, 2:308 (1962).
137. B. H. Alexander, G. C. Kuczynski, and M. H. Dawson, in: Physics of Powder Metallurgy, W. Kingston, ed., McGraw-Hill, New York (1951).
138. A. P. Greenough, Phil. Mag., 43:1075 (1952).
139. F. H. Buttner, H. Udin, and J. Wulff, Trans. AIME, 194:401 (1952).
140. A. L. Pranatis and G. M. Pound, Trans. AIME, 203:664 (1955).
141. E. R. Hayward and A. P. Greenough, J. Inst. Metals (London), 88:217 (1959-1960).
142. L. L. Seigle, in: Progress in Powder Metallurgy, Proc. 20th Ann. Powder Metallurgy Technical Conf., New York (1964), Vol. 20, p. 221.
143. R. A. Andrievskii, V. V. Panichkina, and I. M. Fedorchenko, Metal. i Term. Obrabotka Metal., No. 7, p. 48 (1961).
144. M. Eudier, Symposium on Powder Metallurgy, London, 1954, Spec. Rep. No. 58, The Iron and Steel Inst., London (1956), p. 59.
145. H. Wiemer and R. Hanebuth, Arch. Eisenhüttenw., 3:129 (1949).
146. M. Clasing, Z. Metallk., 49:69 (1958).
147. V. M. Likhtman, E. A. Shchukin, and P. A. Rebinder, Physicochemical Mechanics of Metals [in Russian], Izd. AN SSSR (1962).
148. P. Shaninian and M. R. Achter, Trans. AIME, 215:37 (1959-1960).
149. H. Jones and G. M. Leak, Acta Met., 20:21 (1966).
150. G. C. Kuczynski, L. Abernethy, and J. Allan, Plansee Proc., 35 (1959).
151. W. D. Kingery, ed., Kinetics of High-Temperature Processes, MIT Press (1965).
152. J. G. Early, F. V. Lenel, and G. S. Ansell, Trans. AIME, 230:1641 (1964).
153. G. Grube and L. Schlecht, Z. Elektrochem., 44:367 (1938).
154. W. D. Kingery, J. Appl. Phys., 3:301 (1959).